New Land For Old

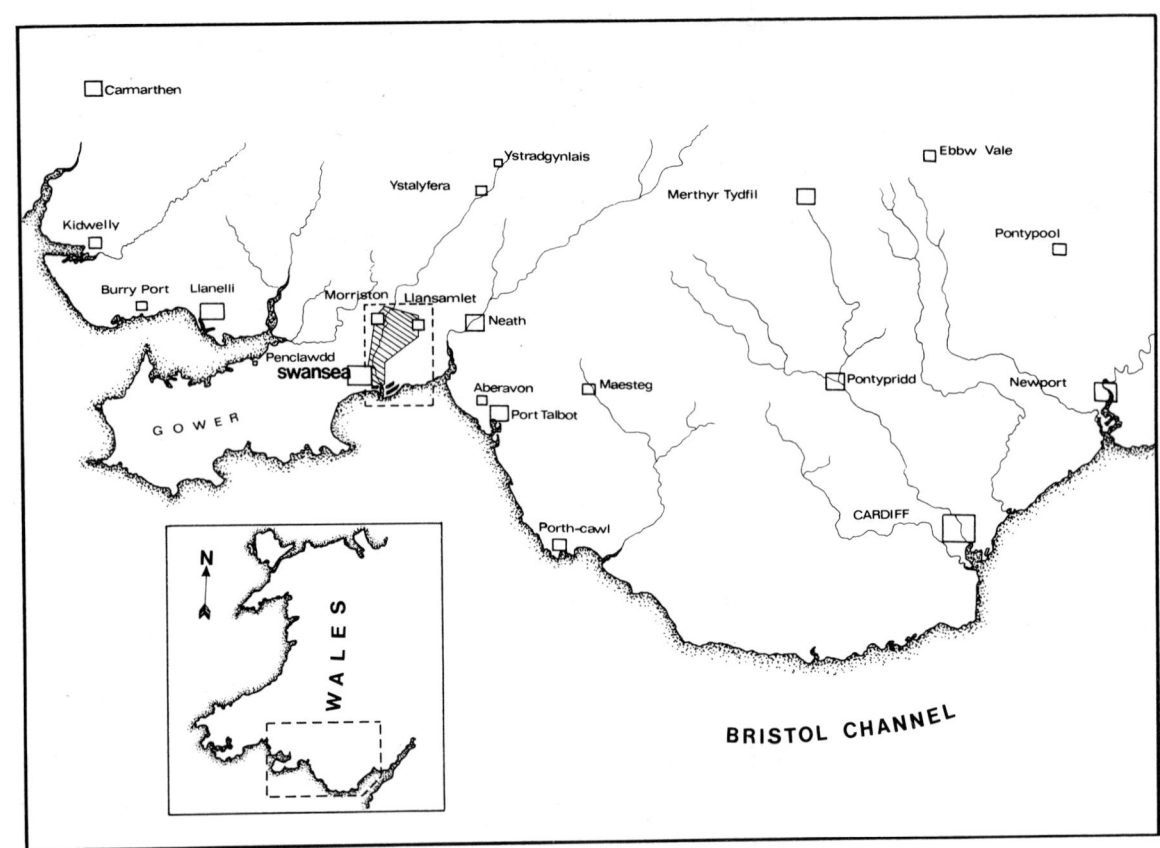

Map of South Wales showing the location of the Lower Swansea Valley.
(CJM 1979)

New Land For Old

The Environmental Renaissance of the Lower Swansea Valley

Stephen J Lavender

University College of Swansea

Adam Hilger Ltd, Bristol

Copyright © 1981 S J Lavender

All rights reserved. No part of this publication may be reproduced, stored in a retrieval system or transmitted in any form or by any means, electronic, photocopying, recording or otherwise, without the prior permission of the publisher.

British Library Cataloguing in Publication Data

   Lavender, Stephen
   New Land For Old: the environmental renaissance of the Lower Swansea Valley.
   1. Reclamation of land—Wales—Swansea Valley
   2. Derelict Land—Wales—Swansea Valley
   I. Title
   333.73′09429′82   HD1671.G7

   ISBN 0-85274-386-6
   ISBN 0-85274-453-6 (Pbk)

Consultant Editor: **Professor D B Ager,** Department of Geology, University College of Swansea.

Published by Adam Hilger Ltd, Techno House, Redcliffe Way, Bristol BS1 6NX.
The Adam Hilger book-publishing imprint is owned by The Institute of Physics.

Typeset by Quadraset Ltd, Radstock, and printed in Great Britain by J W Arrowsmith Ltd, Bristol.

*For Neil
and future generations*

Here nothing lived, not even the leprous growths that feed on rottenness. The gasping pools were choked with ash and crawling muds, sickly white and grey, as if the mountains had vomited the filth of their entrails upon the lands about. High mounds of crushed and powdered rock, great cones of earth fire-blasted and poison-stained, stood like an obscene graveyard in endless rows, slowly revealed in the reluctant light.

They had come to the desolation that lay before Mordor . . .

J R R Tolkien, *Lord of the Rings*
(© 1954 George Allen & Unwin)

Foreword

by HRH The Prince of Wales

For twenty years those responsible for the Lower Swansea Valley project have been carrying out magnificently successful work in rehabilitating industrial wasteland in the valley and making it available again for residential, industrial and recreational use.

I am proud to say that for ten of those years The Prince of Wales' Committee has been associated with this work, in giving Awards for particularly worthwhile schemes, assisting with volunteer projects, and helping to publicise both the work that has been done and the continuing needs of the project.

I am delighted that this remarkable work of environmental renaissance should now have been described in this book by Mr Stephen Lavender, Conservator of the Project. It is an inspiring story and it is surely appropriate that it should have been told by someone so closely associated with it.

Preface

Nowhere in Britain was the problem of derelict and polluted land more acute than in the Lower Swansea Valley. Two hundred years of metal smelting had created a thriving industrial community out of a small village. The industry, however, destroyed the natural vegetation of the area as a direct result of its pollution. When the remaining industries closed during the 1930s a legacy of tips, derelict buildings and bare eroded hillsides was left behind.

Deep resentment built up against the polluters of the valley and this led to the belief that the devastated valley might never recover. In 1961, however, the initiative of one man served to break the apathy: his proposals brought about the establishment of a major project to investigate the problems of the valley. This, in turn, led to suggestions as to the procedure for its complete reclamation. The redevelopment work was carried out, and by 1980 the Lower Swansea Valley had become a major example of how successfully derelict land can be reclaimed.

Throughout Britain whenever coal could be extracted or where metal bearing ores existed, communities developed and reflected Swansea's dramatic growth and prosperity. These areas also followed the same industrial decline and decay. Today many cities are suffering from the problem of derelict land created by neglected and unused industries as Swansea did. Now the Lower Swansea Valley is being redeveloped in a variety of ways and is itself reflecting a change in attitudes towards derelict land.

During the 1970s we have had a Conservation year, a Tree Planting year, an Architectural Heritage year and in 1981 a Campaign for Urban Renaissance. In Wales, the Prince of Wales' committee has been established to provide expertise and financial aid to help promote volunteer conservation projects. In London the Ecological Parks Trust and the recently formed Urban Wildlife Group based in Birmingham have been established to generate interest in city-based projects.

The Lower Swansea Valley has been involved in all the designated 'years' and has provided advice and information for the national bodies. This change of interest in urban areas and the unique history of the valley stimulated the writing of this book. Many schools, colleges and universities visit the valley each year as a first-class study area to demonstrate the problems and techniques of reclamation. It is hoped that this book will give heart to those people who live in cities, and furthermore stimulate people with similar problems into action.

S J Lavender
University College, Swansea

Acknowledgments

I would like to thank the following people and organisations for their help and assistance in the preparation of this book.

    Mr C J Marland for producing the majority of the line illustrations and maps;

    Dr M J Isaac and Miss B Nelmes for their assistance in providing access to archival material at the Swansea Museum;

    Dr S J Wainwright for his criticism and constructive comments on the text;

    Mr H Holloway and Mr M James for their assistance into the research and the production of photographic material;

    Swansea City Council;

    Swansea Museum;

    Mrs J Lavender, Miss J Wells, Miss T E D Williams and Mrs S J Williams for their preparation of the script.

    Finally, I would like to thank my wife Teresa for her patience and encouragement throughout.

The Lower Swansea Valley Project Report, edited by K J Hilton, and published by Longmans in 1967, has been my major source of reference and is quoted in a number of chapters.

The Swansea Valley is covered by the Ordnance Survey 1:50 000 sheet number 159.

Contents

Foreword by **HRH The Prince of Wales**	vii
Preface	ix

1 Swansea's Industrial Revolution — 1

    Early Swansea — 1
    Coal — 2
    Early Copper Smelters — 4
    Development of the Copper Industry — 6
    The Vivian Family — 11
    Further Development of the Copper Industry — 12
    The Zinc Industry — 16
    Transport in the Valley — 19
    The Decline of Copper and Zinc — 27
    Tinplate — 30
    Steel — 31
    Subsidiary Valley Industries — 34
    Social Conditions in the Valley — 35

2 The Environment — 42

    Aerial Pollution — 43
    Early Attempts to Combat Aerial Pollution — 48
    Attitudes to the Valley — 49
    Solid Tip Waste — 51
    Pollution of the River Tawe — 56
    Dereliction — 59
    Early Reclamation Plans — 62

3 The Lower Swansea Valley Project — 68

    The Initiative — 68
    The Project — 69

	Studies of the Project	71
	The Lower Swansea Valley Project Report	76
	Planning in the Valley	78
	Land Acquisition	79

4 Reclamation — **81**

    Early Voluntary Work — 81
    The Photographic Record — 83
    Grant Aid and Major Reclamation Work — 84
    The Conservator and School Involvement — 95
    Community Involvement — 104

5 The Future — **111**

    The Achievements — 111
    Work in Progress — 115
    The Future — 124

Bibliography — **127**

Index — **131**

Chapter 1

Swansea's Industrial Revolution

> ... A very considerable town with a very great trade for coals and culmn, which they export to all parts of Sommerset, Devon and Cornwall, and also to Ireland itself; so that one sometimes sees a hundred sail of ships at a time loading coals here, which greatly enriches the country and particularly this town of Swanzey, which is really a thriving place.
>
> *Daniel Defoe, 1724*

Early Swansea

Throughout South Wales settlements had grown up in the valleys and along the coast long before the Industrial Revolution. One such settlement, on the west bank of the Tawe estuary, was Swansea, where the Normans had built a castle to defend the landward approaches to the fertile Gower Peninsula. The village developed, and by the mid-seventeenth century there were 153 houses clustered around the castle and along the banks of the River Tawe.

The castle nestled between the slopes of Kilvey Hill and Mayhill, clothed in their woodlands of birch and oak. The clear River Tawe, whose source is high in the mountains of the Brecon Beacons, flowed through this green and pleasant valley, by the castle steps, and on into the magnificent Swansea Bay. The inhabitants of this

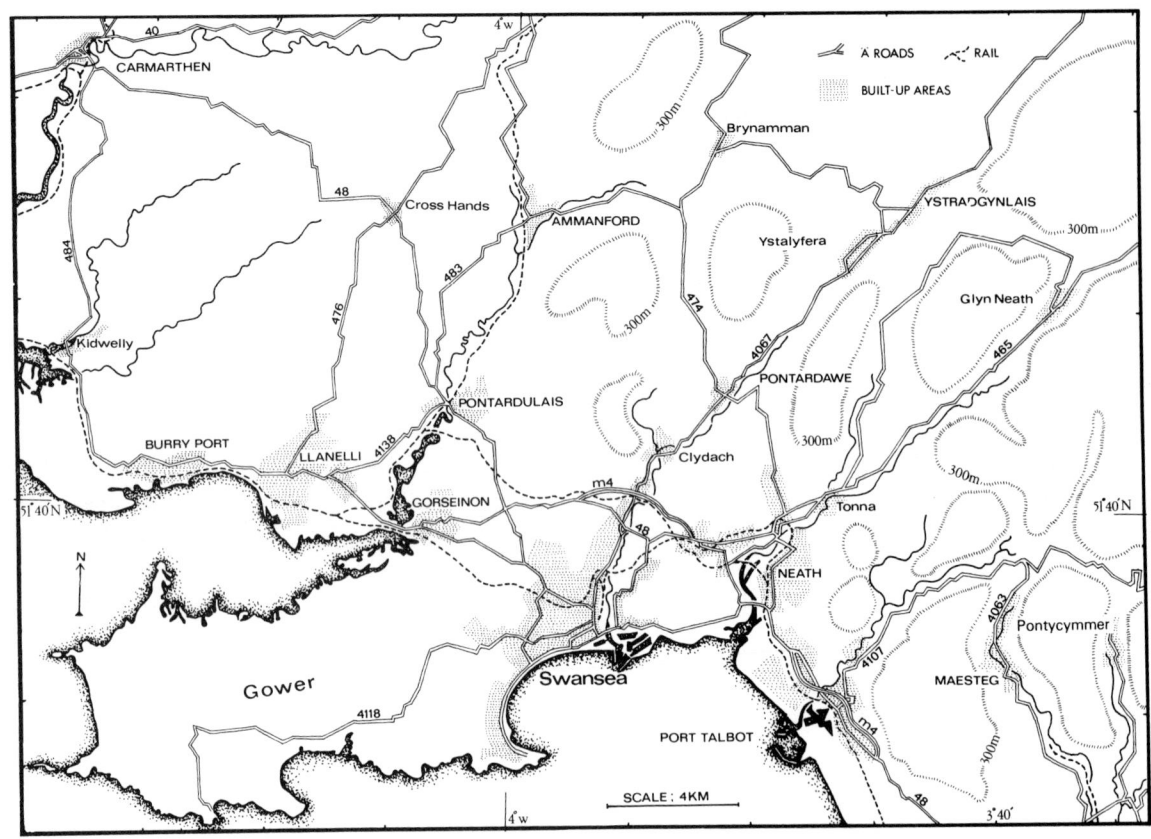

Detailed map of the Swansea area.
(CJM 1980)

small village followed a variety of occupations, particularly farming, fishing and the cockle industry. But their lives and their countryside were to be transformed by the discovery of coal.

Coal

In the thirteenth century coal was already in use in Britain as a domestic fuel. Where the seams were exposed it was a simple matter to work the outcrops and extract the coal. In Swansea the true mining of the outcrops began during the fourteenth century, and by the sixteenth, coal was being exported to Devon, Cornwall, the Channel Islands and Ireland.

View of Swansea Castle across the River Tawe in the eighteenth century.
(From an engraving at Swansea Museum)

Industry was first attracted to the Swansea Valley when the superior quality of the easily accessible local coal proved to be most suitable for the metal ore smelters of the eighteenth century, although there were other historical reasons for the establishment of smelters in the valley.

Prior to the seventeenth century, iron had been smelted from its ores in furnaces fired by charcoal. Charcoal was produced in works built in heavily afforested areas, particularly around the Weald of Kent and Sussex and the Forest of Dean in Gloucestershire. These forests, however, were also the sole source of the timber needed to build ships for the British Navy, and the concern about their rapid depletion resulted in an act, passed under Queen Elizabeth I, which imposed severe restrictions on the felling of trees and the activities of the charcoal burners. From this point charcoal was at a premium for the smelters.

There was then a real need to find a new source of fuel for industry. In 1611 Simon Sturtevant attempted to use coal in place of charcoal, but the sulphur impurities in the coal caused such a deleterious effect on the metal produced that the method was abandoned. Ten years later, in 1621, Dud Dudley of Staffordshire succeeded in producing good-quality cast iron and wrought iron by using coke instead of unprocessed coal, but this method was also abandoned after arousing intense antagonism from the charcoal-burning industry. However, in the early eighteenth century, Abraham Darby of Coalbrookdale finally achieved the acceptance of coal as a fuel for smelting iron. He produced quality iron, and so, together with other improvements, the process of the conversion of iron into steel was revolutionised. After this time the centres of the smelting industries therefore moved from the forests to the coalfields. The value of the Swansea coal was now realised, and it was only a matter of time before this new resource could be extracted on a large scale, and the exploitation of the landscape could begin.

Early Copper Smelters

The first copper ore smelter associated directly with the Lower Swansea Valley was established at Landore in 1717, although other smelters had been located at Aberdulais near Neath during the sixteenth century.

In 1564 an Anglo–German partnership, established under Elizabeth I as the Mines Royal, began mining and smelting copper in the Lake District at Keswick. Ulrich Frosse, who had been involved with the smelting operations there, took over the management of the Mines Royal's affairs at Aberdulais in 1584. Two years later this plant was able to smelt 24 hundredweight (1.2 tonnes) of ore per day with only one furnace. It is probable that this amount of copper could only have been produced using a reverberatory furnace, and if so this is the first example of the use of such a furnace in Britain.

Although it has been established that the first copper smelting was carried out at Neath in 1584, the greatest developments, particularly in smelting techniques, were made in the Swansea Valley itself. John Lane, a Bristol physician, and his partner John Pollard owned copper mines in Cornwall, but decided to build their smelter at Landore on the coalfield rather than at the source of the ore, for a number of reasons. In the first place, the smelting process in use at this time required 18 tonnes of coal to smelt 13 tonnes of copper ore, to produce one tonne of refined copper, so that if the smelters had been built in Cornwall the cargo vessels would have had to carry coal there, but would have had to return in ballast. The ships which transported ore from Cornwall to Swansea, on the other hand, were able to return carrying coal as ballast, which could then be used to raise steam in the pumping engines removing water from the copper mines.

Secondly, because the South Wales coal was very near to the surface and therefore easily extractable, labour and removal costs were relatively low, and consequently its price was only one-third of that mined in other parts of Wales. In addition, the River Tawe provided easy access almost three miles (4.8 km) upstream, and the ore-

Ships in Swansea Bay in about 1700.
(SM)

Llangyfelach copper works at Landore in about 1720, the first copper smelter to be built in the Swansea Valley.
(SM)

carrying ships could discharge their cargoes directly at the smelters, again avoiding heavy transport costs which might have been incurred in getting the ore from the ships to the smelters overland.

Lane's Llangyfelach works, built at Landore in 1717, consisted of two double houses, a round house, twenty furnaces for smelting the ores, a refinery house, accounting house, a smith's forge, laboratory, test house, rod mill, blast house and two other refining houses. In 1726 the Llangyfelach works were taken over by the Morris family after Dr Lane and his associates went bankrupt. Smelting continued at these works, and gradually more and more smelters were attracted to Swansea.

Development of the Copper Industry

> It is our unanimous opinion that . . . the said works will prove very much to the advantage and not in the least prejudicial or hurtful to our said Borough and inhabitants thereof.
>
> *Alderman G Powell, 1720*

In the eighteenth century the local aldermen were keen to see the copper-smelting industry become established in

View over the village and harbour of Swansea in the early nineteenth century.
(J G Wood, 1813)

Swansea so that the town would grow and prosper.

By this time developments had taken place to improve the steam engine which could be used to raise water, and by 1711 such engines were being installed to pump water from coal mines. In the Swansea Valley, pumping engines were installed so that coal could be extracted from less accessible and deeper coal seams. As further improvements were made, the steam engine was adapted to provide ventilation to the shafts.

Following the establishment of the first smelter at Landore so other smelters were also built in the valley. In 1720 the Cambrian works was set up near the mouth of the River Tawe on the west bank, and this continued to produce refined copper until 1745. The site of this plant was then occupied by the Cambrian Pottery, which produced china and porcelain of the finest quality until it too eventually closed down in 1870. (Many examples of the fine porcelain produced at the Cambrian pottery can now be seen in Swansea Museum.)

The White Rock copper works was established on the site of an old flour mill at Pentrechwyth in 1737, and copper was to be smelted there for more than 150 years. When the works was eventually taken over by the Vivian family in the early 1870s, silver and lead were also being produced.

At the northern end of the Lower Swansea Valley, the Forest copper works was set up in 1746, Chauncey Townsend's Middle Bank works was established in 1755,

The Cambrian Pottery in 1791, built on the site of an old copper works.
(SM)

and two years later the Upper Bank works was opened. Both of the latter smelted lead and copper, but it has been reported that Townsend also produced zinc at his Upper Bank works as early as 1757. The Rose copper works, built in 1780, was followed in 1793 by the Landore works and by the Birmingham or Ynys smelter in the same year.

White Rock copper works on the banks of the River Tawe.
(J G Wood, 1813)

By 1800, nine copper smelters were in production in the valley, and as the fortunes of their owners altered, so the works changed hands, even though the sites and smelting procedures remained essentially the same.

In 1810 the world output of smelted copper was about 9200 tonnes per annum. The Swansea smelters were generally small in size and each afforded employment to between 30 and 70 workers. However, in 1810 the Vivian family arrived from Cornwall and built the large Hafod copper works which eventually employed over 300 men. By 1823 the valley's copper works, together with the collieries and the shipping trade dependent upon them, supported a population in Swansea of 8–10 000. The

The Vivians' Hafod copper works in 1812.
(G R Edwards)

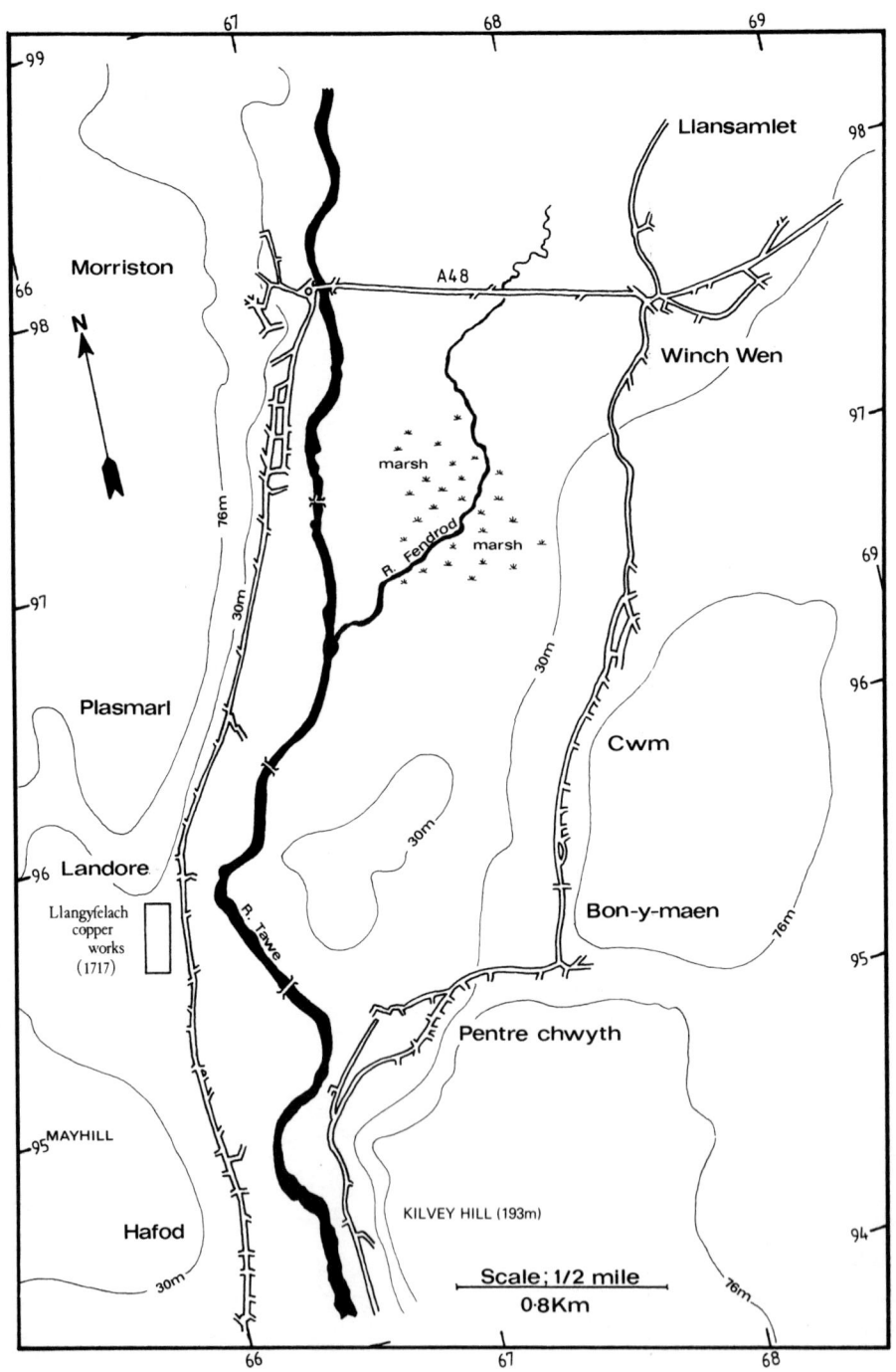

Map of the Lower Swansea Valley showing the main villages, and the site of the first copper smelter at Landore.

developments which took place following the building of the Hafod works helped to ensure the prosperity of the city—but the price that would have to be paid for this prosperity was unforeseen by both the entrepreneurs and the employees of the copper industry.

The Vivian Family

John Vivian and his family arrived in Swansea in the early years of the nineteenth century, and first became involved in copper smelting to the west of Swansea at Penclawdd on the Burry Estuary. In 1810 he transferred his interests to the Swansea Valley, where he built the Hafod copper works on land leased from the Earl of Jersey and the Duke of Beaufort in the names of his two sons, John Henry and Richard Hussey Vivian. The smelter was built on a site between the River Tawe and the recently completed Swansea Canal.

John Henry Vivian's family of nine included four sons, one of whom, Henry Hussey Vivian, took over as manager of the Hafod works when he was only 24 years old, upon the death of his father in 1855. At this time Britain was the largest producer of smelted copper in the world, but as the richer veins of Cornish ore began to be worked out, so it became necessary for the smelters to import ores from other parts of the world. The first foreign ores arrived at the Hafod works in 1827 and, with the accompanying increase in copper production, larger docks were soon required to cope with the larger volume of shipping.

At the Hafod works copper smelting continued, and many other metals were also extracted from the copper-bearing rocks which were imported into Swansea. A number of patents were taken out for the production of cobalt, nickel, silver and gold; by 1864 sulphuric acid was being commercially produced, and in 1871 Henry Hussey extended the operations at White Rock to treat silver and lead ores.

Henry Hussey Vivian. (SM)

The Hafod and Middle Bank works in 1860.
(From *Le Tour du Monde*, Hachette, 1865)

The Vivian family became very rich and were thus able to exert a considerable influence on the development of Swansea. They built themselves several fine homes in the west of the city.

Further Development of the Copper Industry

Williams Foster and Company set up a copper smelter adjacent to the Vivians' Hafod works in 1835. Such was the degree of secrecy concerning the finer details of copper extraction that a huge wall was built dividing the two concerns, and workers were sworn never to disclose the details of the processes used in the works to anyone except their own children. The Landore works was established to produce silver between 1853 and 1855, followed by the Little Landore copper works in 1863. The final smelter to start production in the valley was the Llansamlet copper and arsenic works in 1866.

As the processes within the smelters became more scientific, so the extraction of copper metal from its ores became more objective and efficient. Copper ores arriving at the various Swansea docks were unloaded and

Copper pan being hammered out by steam-hammer.
(SM)

chemists from the different works carried out sampling and analysis procedures to test for quality. Ores were then sold to the highest bidder by auction—a method known as 'ticketing'.

Following a sale, ores were transported up the river on ships or barges for unloading at the private wharves of the smelters. Once in the works the rocks were hammered into pieces and sorted by girls who carefully separated out the valuable metal-bearing ores. At Hafod an elaborate rail track system was used to transfer the ores to the furnaces. The coking and non-coking coals available from the Swansea Valley pits were mixed in the proportion 1:2. The 'calciner' used was a reverberatory furnace taking 7 tonnes of ore at a time. In 1848 the 'Welsh process' of copper smelting consisted of the following stages (from Alexander 1955):

(1) Approximately 3–4 tonnes of ore were charged and heated for 12–24 hours and virtually roasted in the calcinating furnace. On completion the material was granulated by quenching in water. By this operation the sulphur content of the ore could be reduced from 31% to 16%.

(2) The calcinated ore was transferred to another furnace and melted with metal slag from stage (4) to give a 'regulus' or coarse metal which contained approximately 35% copper, together with about 35% iron and 30% sulphur. At this point a certain amount of ore furnace slag produced as a residue contained only about 0.5% copper, and this was dumped on the slag bank outside the smelter.

(3) A further calcination of the regulus from stage (2) was carried out with free access to air for 24 hours. During this stage a considerable amount of sulphur was evolved, reducing its content in the regulus from 30% to 15%, and the oxygen content was raised to 11%.

(4) The partially oxidised coarse metal was resmelted with further additions of oxide to produce a white metal. The white metal now contained 75% copper and 21% sulphur. The oxygen originally present in the charge was used to eliminate iron sulphide in the slag. The slag from this operation contained approximately 34% silica, 56% ferrous oxide, and 5–7% mixed oxides. The white metal was then cast into the form of pigs.

Copper casting by hand from a reverberatory furnace in Morfa. Plain cast-iron moulds were used; the castings were then heated and rolled into sheets and plate. (SM)

(5) The white metal pigs were now charged in a melting furnace and heated very slowly with free access to oxygen so that melting took 6–8 hours. This operation resulted in the formation of blister copper, comprising 98.4% copper, 0.7% iron, 0.3% nickel, cobalt and manganese, 0.4% tin and arsenic, and 0.2% sulphur. The slag was essentially siliceous, containing 47.5% silica, 17% cupric oxide, 2% copper, 28% ferrous oxide and 3% alumina. This slag was then recycled in stage (3).

(6) The final operation was refining, resulting in marketable copper, and a refining slag which was recycled. In the refinery furnace 6–8 tonnes of blister copper were melted and exposed to air for 15 hours to oxidise; the slag was then skimmed and dry copper made. The charge was then covered with anthracite or a free-burning coal, and a thick birch or oak pole—the greener the better—was placed under the surface of the molten metal. This poling operation resulted in the evolution of considerable steam and free hydrogen, which eliminated the bulk of the oxygen from the molten metal. Sampling at this point of the appearance of the

Warehouse for copper circles and sheets. Note the destination of the sheets in the foreground—Bombay.
(SM)

surface 'set' of small button samples cast in chilled iron moulds determined the satisfactory completion of the operation.

Essentially, a satisfactory copper with good malleability usually had a flat, wrinkled surface and, when bent double after hammering, had a silky fracture. This copper was known as 'tough pitch' and at this point would be ready for casting. The metal was ladled by hand, 30 lb (13.6 kg) at a time, into suitable moulds for solidification as cake, billet or strip.

In *The Principles of Copper Smelting,* published in 1907, E D Peters reported that at least one American professor of metallurgy believed that the Welsh process of copper smelting was 'the most beautiful of metallurgical processes'. Peters also described the process thus:

> The Swansea smelters developed great skill in the construction and management of the reverberatory, and found it particularly suitable to the great variety of finely pulverised ores of every conceivable composition, which reached their port from all parts of the world.

Swansea's copper smelters reached their peak around 1860 when the works there were smelting two-thirds of the copper ores imported into Britain. The world production of copper had now climbed to over 300 000 tonnes per annum.

The Zinc Industry

The copper industry in the Swansea Valley started to decline during the 1870s, and it was fortunate for the working population that their smelting skills and the valley's industrial location were ideal for the smelting of other metals. Zinc (or spelter) was first produced in its metallic form in Britain in 1738 by William Champion, a former copper smelter from Bristol. It has already been mentioned that Chauncey Townsend is recorded as having smelted zinc at his Upper Bank works in 1757, but there is no further evidence of zinc smelting in the valley until 1836.

During this time many advances were being made both in the smelting processes and in the use of zinc. In 1780 brass was produced as a direct alloy of copper and zinc, and later (1805) it was discovered how to roll zinc into a sheet. George Frederick Muntz invented another copper–zinc alloy in 1832, which was eventually used to replace copper sheathing and accessories on ships.

The Muntz alloy was followed in 1837 by the development of a process of coating iron sheets with zinc. This 'galvanising' process revolutionised the demand for zinc, and galvanised sheets were exported in their millions throughout the world. The world production of zinc thus rose from less than 5000 tonnes in 1830 to about 350 000 tonnes in 1890.

The Cambrian spelter works, owned by Evan John, was the first successful zinc smelter in the Swansea Valley, and was opened in 1836. In 1841 the Vivian family extended their industrial empire by converting the

Zinc distillation furnaces at the Upper Bank works.
(SM)

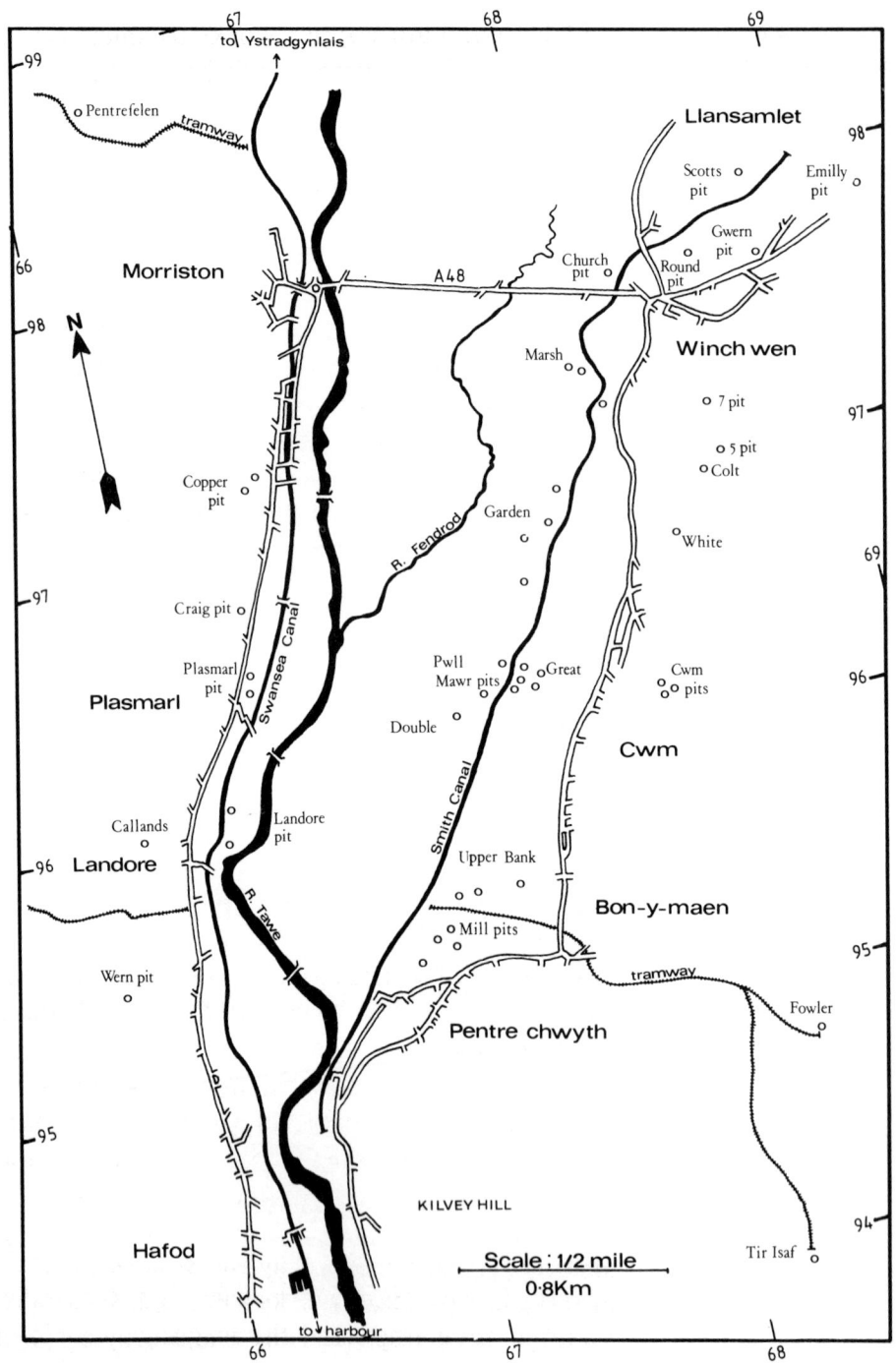

Map showing the main coal mines in the valley, and the courses of Smith's Canal and the lower section of the Swansea Canal.
(CJM 1979)

old Birmingham copper works to smelt zinc ores, and other zinc plants were established by the Villiers family in 1873, followed by the Swansea Vale works in 1876, the Glamorgan works in 1887, and the Upper Bank works for a short time between 1924 and 1930.

Working conditions in the zinc smelters were extremely harsh, and a high degree of furnace management was required. At certain times Belgian and German workers were employed there, both because of their familiarity with smelting furnaces, and also because of the occasional difficulty in obtaining local labour to work in such harsh conditions.

The eventual decline of the zinc industry followed a very similar pattern to that of the copper industry, and is discussed later in this chapter.

Transport in the Valley

As the industries developed, a considerable transport system was established, with tracks, wagonways, canals, docks and railways. Each improvement contributed to the complete industrialisation of the Lower Swansea Valley.

Chauncey Townsend, an alderman of the City of London, secured an interest in the Church Pit coal mine in Llansamlet which had been developed by his son-in-law, John Smith. Townsend built a wagonway to link his smelters with the pits in Llansamlet, which was simply a track about three miles (4.8 km) in length along which horse-drawn carriages could carry their loads of coal. In 1757 Townsend and Smith established the Upper Bank smelter, and the wagonway was able to supply both this and the Middle Bank works with coal. John Smith's son, Charles, later improved the wagonway by laying down iron rails. The horse-drawn carriage was found to be a much more efficient form of transport if the horse had a

The Swansea Canal at Hafod. (SM)

definite line to follow, rather than simply being pulled along a track.

In the late eighteenth century, however, the relatively inexpensive form of transport by barge was being developed, and in 1784 the wagonway was completely replaced by a canal known locally as Smith's Canal. It too was only three miles long, without a single lock, and was used exclusively for the transport of coal from Smith's collieries to Townsend's smelters.

The development of the canal network throughout Britain was stimulated by the growth of industrial centres during the late eighteenth century, especially since the roads were often very poor, and because of the rapid increase in the demand for cheap ways of moving goods. The first major canal to be built in Britain was the Bridgewater Canal, completed in 1761, to carry coal from Worsley to Manchester, and its success initiated an intense programme of canal building. A number of inland towns, particularly those in the Midlands, were thus able to expand as a result of these new trading routes. During the boom of the early nineteenth century, over 4000 miles of navigable waterways were built, but the expansion was to end suddenly in the 1830s with the growth of the railway system.

As the coal consumption of the smelters in the Lower Swansea Valley increased, so the mines immediately surrounding these works inevitably became worked out, making it necessary for coal to be brought from mines further up the valley. In order to offset the transport costs which inevitably rose, it was decided that another canal should be built to run parallel to the west bank of the River Tawe, to link the smelters with these mines. The building of the Swansea Canal took four years (1794–98), and was associated with the remodelling of the harbour. An Act for the building of the canal was introduced in 1794, and the canal company set about obtaining the permission of the landowners and occupiers for the canal to cross their land. It ran from the

Map showing the various stages in the development of the Swansea docks, and the positions of the castle and the Swansea Museum/ Royal Institution of South Wales. (CJM 1979)

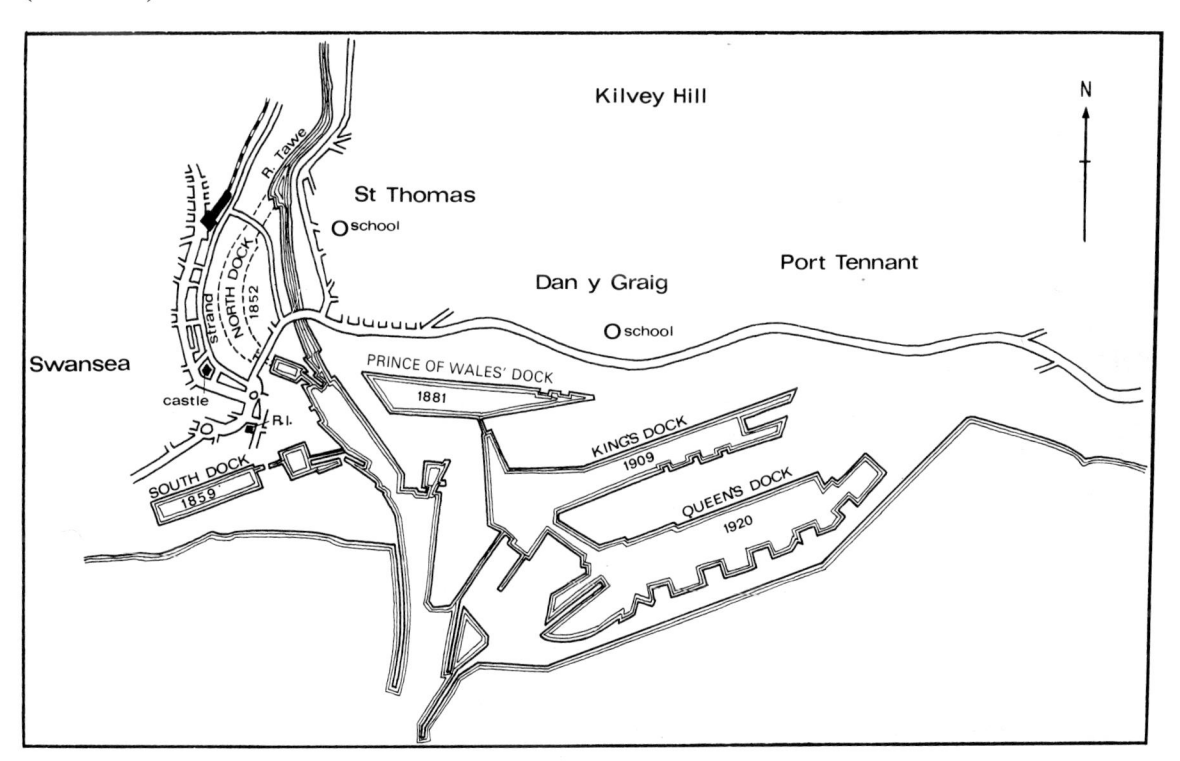

Strand in Swansea, to Ystradgynlais in the north of the county of Glamorgan, extending a total length of just over 16 miles (25 km), rising 400 feet (120 metres) and with 36 locks. Once opened, the canal was able to maintain the supply of coal and to ensure the prosperity of the industries and inhabitants of Swansea.

With the increase in shipping to serve the needs of the valley's industries, so improved port facilities were required to handle the additional traffic. In 1791 an Act was passed which founded a group of elected members—the Trustees—as the Swansea Harbour Trust. It was the responsibility of the Trustees to repair, enlarge and maintain the harbour, although the port did not begin to take its present shape until the early nineteenth century.

In 1852 the North Dock was created by enclosing a meander of the River Tawe and digging a new channel

View of the Lower Swansea Valley looking north in about 1885, showing the development of the city and the harbour, and the pollution produced by the smelters.
(SM)

22

the New Cut—for the river. A second dock on the west bank of the river mouth was authorised in 1847; the land was acquired by the Trustees in 1857, and the South Dock was opened in 1859. The first dock on the east bank of the river, the Prince of Wales' Dock, was opened in 1881, and this extended over the site of a canal basin complex in 1898. This was followed by the building of the King's Dock in 1909, and Queen's Dock in 1920.

Most of the traffic using the dock facilities in the nineteenth century consisted of clippers and barques. The clippers became very important between 1830 and 1870 as they were able to carry their cargoes at high speed. Designed for speed with three masts and a slim line, the clipper had a sheer curve from bow to stern. The skills of the crews of these ships were vitally important since every change in wind direction meant that all hands were required to alter the sails.

With the need to transport heavy, bulky cargoes of ores over long distances, larger vesesels were required, so the slim lines of the clipper were redesigned to produce a longer, fuller-bodied ship. It also became necessary to add a fourth mast in some cases, and this new style of ship was called a barque. The barque was considered to be the most satisfactory ship for transporting copper

Copper ore barques in Swansea docks.
(SM)

The brig L'Espérance *leaving Swansea for the west coast of America via Cape Horn.*
(SM)

ores from the mines all over the world to the smelters on the River Tawe.

By the year 1901 copper ores were being imported from Chile, Mexico, Norway, Spain, Portugal, Newfoundland, Africa, Ireland and Australia, as well as Cornwall. In addition, copper bars and cake were being imported from the USA, Chile, Turkey and East Africa for further processing.

At about the same time as the redevelopment of the Swansea docks was under way, came the railway. Before 1847 there had been no direct rail link with Swansea over the River Tawe, but in that year work was started on the Landore Viaduct, designed and built by Isambard Kingdom Brunel. The expansion of the railway system was rapid following the opening of the Liverpool to Manchester line in 1829, after Stephenson's *Rocket* won a competition by drawing a twenty-ton train 35 miles (56 km) in just under two hours. In 1840 the London to Southampton line was completed, and the Great

Isambard Kingdom Brunel.
(SM)

24

Western Railway from Paddington to Bristol was opened in 1841. It then became fashionable for investors to put their money into railways, and between 1844 and 1847 railway tracks were laid all over Britain, although not to any coherent plan.

Brunel was working on improvements to the Bristol docks when he became involved with the promoters of the Great Western Railway, and in 1833, when he was only 27 years old, he became chief engineer of that company. By 1847 his work had brought him to South Wales, where the line from Paddington had recently been extended from Bristol to Cardiff.

The bustling prosperity of the docks area in the early 1900s.
(G R Edwards)

The first train to arrive at Landore Station, Swansea, in 1850.
(SM)

The line to Swansea had to cross the River Tawe at Landore, and for this Brunel designed his wooden viaduct which took three years to build and was opened in 1850. In 1852 the Swansea Vale Railway Company built a line from the docks up to the east bank of the river, alongside Smith's Canal and then on to collieries further up the valley. In 1871 the same company built a line from the Upper Bank works to Morriston, and in

The first Garrett articulated locomotive of this size in Britain, which replaced two engines previously used and could climb severe gradients and negotiate sharp curves with ease. The engine sheds of Vivian and Sons, and the chimneys of the Hafod works can be seen in the background.
(SM)

The Landore Viaduct after its opening in 1850, and horse-drawn barges on the Swansea Canal. (SM)

1881 established another from Morriston to Landore down the western side of the river. Thus the valley was divided into six sections by the river, canals and railway lines; this fragmentation was to cause considerable difficulties in the planning for the redevelopment of the valley in years to come.

The Decline of Copper and Zinc

Copper smelting reached its peak in the Swansea Valley during the 1860s, and although large amounts of copper ore were imported up to 1890, the industry had already begun to decline. Bessemer had recently developed a new and more economical process for copper refining, but its immediate possibilities were not taken advantage of in Britain. The valley's industrialists resisted any suggestions to convert their smelters from the well tried Welsh process to new methods, and this stubborn resistance to change was to be a major factor in the decline of the South Wales copper industry.

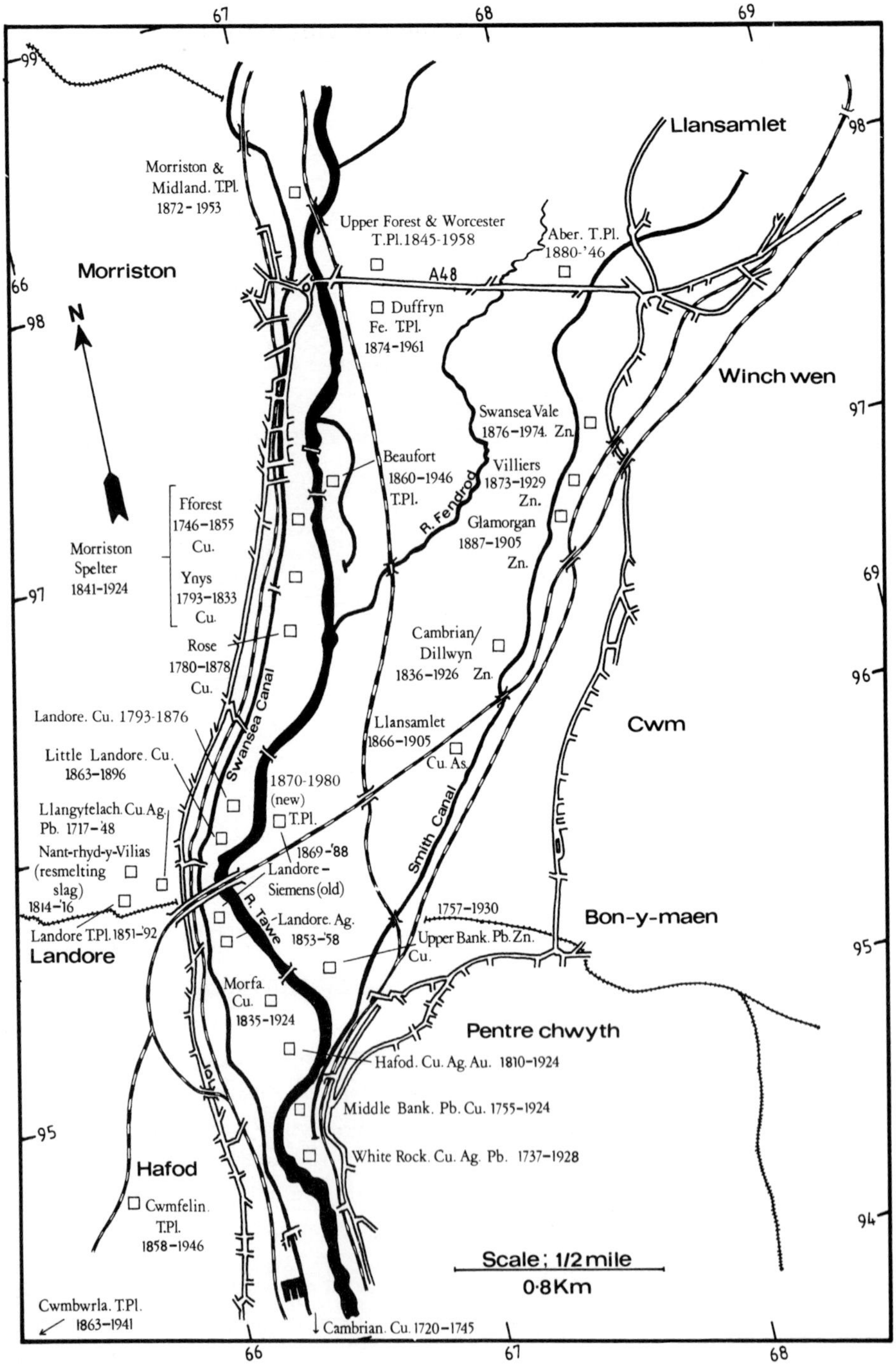
Llansamlet
Morriston & Midland. T.Pl. 1872-1953
Upper Forest & Worcester T.Pl. 1845-1958
Aber. T.Pl. 1880-'46
Morriston
A48
Duffryn Fe. T.Pl. 1874-1961
N
Winch wen
Swansea Vale 1876-1974. Zn.
Beaufort 1860-1946 T.Pl.
Villiers 1873-1929 Zn.
R. Fendrod
Glamorgan 1887-1905 Zn.
Fforest 1746-1855 Cu.
Morriston Spelter 1841-1924
Ynys 1793-1833 Cu.
Cambrian/ Dillwyn 1836-1926 Zn.
Rose 1780-1878 Cu.
Landore. Cu. 1793-1876
Llansamlet 1866-1905 Cu. As.
Cwm
Little Landore. Cu. 1863-1896
Swansea Canal
1870-1980 (new) T.Pl.
Llangyfelach. Cu. Ag. Pb. 1717-'48
1869-'88
Nant-rhyd-y-Vilias (resmelting slag) 1814-'16
Landore - Siemens (old)
Smith Canal
Landore T.Pl. 1851-'92
R. Tawe
1757-1930
Landore
Landore. Ag. 1853-'58
Upper Bank. Pb. Zn. Cu.
Bon-y-maen
Morfa. Cu. 1835-1924
Pentre chwyth
Hafod. Cu. Ag. Au. 1810-1924
Middle Bank. Pb. Cu. 1755-1924
White Rock. Cu. Ag. Pb. 1737-1928
Hafod
Cwmfelin. T.Pl. 1858-1946
Scale; 1/2 mile
0·8Km
Cwmbwrla. T.Pl. 1863-1941
Cambrian. Cu. 1720-1745
67
68
69
99
98
97
96
95
94
66

Map showing the location of industries in the Lower Swansea Valley, the dates they operated, and the metals they produced. (T.Pl. = tinplate, Fe = iron, Zn = zinc, Cu = copper, As = arsenic, Pb = lead, Ag = silver.)
(CJM 1979)

As the immense value of copper to the electrical industry came to be appreciated in the early years of the twentieth century, so other countries started to build their own smelters incorporating the newer, more economical techniques, and this put the Swansea works at a considerable disadvantage. The world production of refined copper had soared from 9200 tonnes in 1810 to over 350 000 tonnes in 1898, and as the ore fields of Britain began to be worked out, it became obvious that the Swansea smelters would never be able to compete on a large scale.

After 1890, as cheaper refined copper was imported into Britain, a series of amalgamations gradually brought the copper industry of the valley under the control of the Vivian family and the Williams Foster Company. In 1924 both of these firms were acquired by the Bristol Copper Manufactories Limited, and this in turn by Imperial Chemical Industries in 1928. This type of amalgamation of small companies was later to be repeated in other industries such as those of tinplate and steel.

The zinc-refining industry, which was only just beginning to be developed when copper production was at its height, gradually became more important in the Swansea Valley. Between 1860 and 1914, fourteen copper smelters in South Wales closed down, and eleven of these started to smelt zinc ores. Many copper workers were able to adapt their skills to zinc smelting, and so continued in employment. By the late nineteenth century, Swansea had become the centre of the zinc-smelting industry, producing 20% of the total national output. Once again, however, the industrialists' resistance to change contributed to the decline of zinc refining in the valley.

After World War I new electrolytic processes for refining zinc were introduced and when more efficient modern plants were built close to the supplies of ores in America, Canada and Australia, they proved to be far more able to meet the European demand for zinc, and so sealed the fate of the industry in Swansea.

Between 1924 and 1928, four of the five remaining smelters in the valley closed down. The Swansea Vale works had been modernised in 1916, with government assistance, and in 1960 became the first commercial blast furnace plant in the world producing zinc, although it too finally closed down in 1974.

Tinplate

The tinplate industry became established in Swansea using expertise gained and developments made over the years in other parts of Wales. The first iron bar to be rolled and coated with tin in Britain was produced at Pontypool in around 1660, but the industry gradually moved westwards until by 1880 there were 64 plants in South Wales, 40 of which were to the west of Port Talbot. The Swansea valley's first tinplate works was built at Upper Forest in 1845, and this was followed by ten others built between 1850 and 1880: at Landore (1851), Cwmfelin (1858), Beaufort (1860), Cwmbwrla (1863), Worcester (1868), Morriston (1872), Duffryn (1874), Midland (1879), Birchgrove and Aber (1880).

Following the decline of copper and zinc smelting in the valley, the tinplate industry expanded rapidly in the years preceding World War I, so that by 1913 there were 106 mills in the eleven works. Many workers had acquired useful skills in the copper and zinc smelters, and so this labour force was readily absorbed by the tinplate works. Morriston, which was already a thriving community, developed and became known locally as the 'tinman's suburb'.

Each tinplate mill had three departments: the furnace, in which the pig iron was reheated to produce iron suitable for rolling; the rolls, where the iron bar was rolled out; and the tin house, where the sheets were coated with a layer of tin. In the works mill teams were established, comprising furnacemen, rollermen, doublers, behinders and first and second helpers. Members of these teams were often associated with each other through family and church, as well as work.

In 1887 the Swansea Royal Jubilee Metal Exchange was established, and by 1890 Swansea had become the business centre of the tinplate trade in Britain. In 1891, however, the United States government imposed a tariff on all tinplate imported into the USA, and although this did not seriously affect production in South Wales up to World War I, it did stimulate the United States to develop its own tinplate industry. With the outbreak of war in 1914, the USA was able to step in and fulfil British export commitments to Europe while Britain's own industrial efforts were directed to the war.

Following the war, the old problem of resistance to change arose. While the new process of hot strip milling was introduced in Kentucky in the USA in 1923, the Swansea Valley industrialists continued to use uneconomic methods, which again put them at a competitive disadvantage. Many of the mills in the valley started to close, and others were acquired by Richard Thomas and Company. The aim of this company was to build a fresh structure of larger units, set up to gain sufficient output capacity to build a strip mill. With the opening of strip mills in other parts of South Wales—at Ebbw Vale in 1938 and at Port Talbot in 1947—so all the valley works ceased production. The majority of these mills closed down during the 1940s and 1950s; the last to close was the Duffryn works in 1961.

Steel

During the nineteenth century the demand for iron rose dramatically with the development of the railway system throughout Britain. This put a heavy strain on the production of wrought iron, the only suitable material available at the time, until in 1856 Bessemer invented a new process for making cheap steel in bulk. The Bessemer converter enabled the unwanted constituents in the molten pig iron to be removed much more efficiently, and this revolutionised the steel-making industry. At the same time, C W Siemens and his brother were using similar techniques to develop the open-hearth process, which made more efficient use of the heat generated during the steel-making process, and thus reduced the amount of coal required.

This regenerative open-hearth furnace, which was patented in 1857, used firebricks to preheat both fuel gas and air with the hot combustion products, and the heat produced by the furnace was sufficient to keep the metal molten. This style of furnace was exploited for reheating metals, puddling iron (the conversion of pig iron to wrought iron by stirring in a molten state), melting glass and for producing crucible steel.

William Siemens. (SM)

In 1867 William Siemens and his colleagues formed a private steel-making company, and in 1868 they acquired the Landore silver works. Following the conversion of the site, a commercial open-hearth steel works was built by the Landore Siemens Steel Company, and in 1869 the operation began with an output of around 75 tonnes of steel per week. By 1870 more than 100 tonnes of steel were being produced per week, and in 1871 a larger site was acquired further upstream on the east bank of the river for the building of three modern blast furnaces using the Siemens design. By 1873 the works was turning out 1000 tonnes of steel per week, and had become the fourth largest producer in the world. The works was also a major employer in the valley, with a workforce of 2000 men in 1874.

The Siemens open-hearth process made it possible to produce finest quality steel from pig iron; throughout Britain at the turn of the century more steel was being produced using this method than with the Bessemer converter. From 1875 onwards, however, in spite of the

The unusual chimney stacks of Siemens' 'old' laboratory, now derelict. (LSVP 1963)

technical success of this process, labour costs increased, financial losses were incurred, and the Landore Siemens Steel Company was forced out of business in 1888.

In Swansea the open-hearth furnace was of special value to the tinplate industry, and many were built in the tinplate works.

The steel and tinplate works were very closely interdependent, but as industrial decline set in following the end of World War I, many small firms amalgamated or were bought out by larger companies, and the steel industry in South Wales eventually became concentrated in the more modern large-scale plants at Port Talbot and Newport. After the closure of the Siemens steel works, a portion of the site was acquired by the Mannesmann Tube Company which had been founded in 1884. The Landore Siemens Steel Company then acquired British patent rights for the production of seamless steel tubes using the Mannesmann process. The Siemens family continued to be the major shareholder in this enterprise until 1899, when the Mannesmann brothers, Max and

The Mannesmann steel tube works, later acquired by Stewart and Lloyds Ltd.
(LSVP 1961)

33

Reinhard, acquired a majority share and formed the British Mannesmann Tube Company.

During the four years of World War I, the company was controlled by the Custodian of Enemy Property, but was handed back to the German brothers in 1918, and by 1919 the plant was producing 35 000 tonnes of steel tubing per annum, employing over 1500 workers. In 1938 the company was acquired by Stewart and Lloyds Ltd, but the industry was in the unfortunate position of being served only by rail, with no road or canal access, and this was a major reason for the eventual closure of the tube works in 1960.

Siemens laboratory in Landore, with the abandoned Swansea Canal in the foreground.
(SJL 1978)

Subsidiary Valley Industries

The major non-ferrous metal smelters in the Swansea Valley were also involved in the manufacture of materials of benefit to other industries in the area. Copper and zinc ores often contained other valuable metals which could be extracted economically, and so many of the smelters produced small amounts of these metals. It has already been noted that cobalt, nickel, silver and gold were extracted from ores at the Hafod works; lead and silver were smelted at White Rock; and arsenic was produced at the Llansamlet works.

A number of attempts were made to resmelt waste from the works after it had been realised that early smelting techniques had been very inefficient and that some of the tips might contain relatively large amounts of valuable metals (after all, some of the original copper pyrites ores from Spain, Portugal and Norway contained only 3–5% copper). This resmelting of waste took place at the Nant-rhyd-y-Vilias works in Landore in as early as 1814, and continued in other smelters until as recently as 1952, although there is no record that this operation has ever been commercially viable in the Lower Swansea Valley itself.

The sulphur generated during the smelting of ores was extracted and used to produce copper sulphate, another by-product, which was used as a fungicide. Sulphuric acid was also manufactured in many of the works, and was used in a variety of industrial processes as well as being sold on a commercial basis. The acid was used to produce phosphates for use as fertilisers, and in the tinplate industry to prepare the iron sheets to receive the coating of tin, a process called 'pickling'. Other by-products included zinc chloride, used as a flux in the tinplating process, and zinc oxide, which was used in the pharmaceutical industry.

Social Conditions in the Valley

Inevitably the availability of employment in the Swansea Valley was accompanied by a rapid increase in the working population of the area. In 1700 there were only a few hundred people living in the village of Swansea, but by 1851 the population of the borough had risen to over 21 000.

Around each smelter separate communities became established, and these gradually expanded and coalesced to form the city of Swansea. For example, on the east bank of the river, Pentrechwyth developed alongside the White Rock, Upper Bank and Middle Bank works; and further north, Llansamlet began to grow around the Smith family's coal mines. On the west bank, Landore was already becoming organised into a community, but one of the most important social developments for workers was probably the 'castle' built by the Morris family. In about 1760 John Morris, the younger son of Robert Morris the industrialist, built a large castellated mansion overlooking his smelters on the west bank of the River Tawe, using local stone. Morris Castle had a large central courtyard around which was built accommodation for colliers at the Treboeth pits, and had enough room for forty families; this building was one of the first

The ruins of Morris Castle.
(CJM)

examples of multi-storey living in British history. Accommodation was also provided for a tailor and a shoemaker to cater for the needs of the colliers.

As Morris developed his smelting interests at Landore and Forest, it became necessary to provide further housing for his workers. Together with William Edwards (who had previously built bridges at Pontypridd, Pontardawe, Aberavon and Wychtree) as his architect, he established a village in 1768, which expanded over the years and today still bears his name—Morriston. The village was originally built on a grid-iron pattern, and each house was provided with sufficient land to support a cow. By 1796 Morriston had developed into a small township of 141 houses with a population of 619.

With the building of the massive Hafod works in 1810 it again became necessary to provide accommodation for the extra workers and their families. The village of Hafod (or Trevivian, as it was often called), developed in a very similar way to Morriston. The small terraced cottages, often built of stone and cast slag, consisted of two rooms upstairs and two down, while the larger cottages had two spacious rooms, a parlour, kitchen and a passage downstairs, and three rooms upstairs. Behind each house lay a strip of garden containing a privvy, a pigsty and a coal-hole. The workers were given an

An example of the use of cast slag as coping stones in Hafod. (SJL)

Wychtree Bridge, Morriston (architect, William Edwards) and the Upper Swansea Valley. (J G Wood, 1813)

View over Morriston, the tinman's suburb, in 1965. The Tabernacle Church is clearly visible, with the derelict tinplate works beyond.
(LSVP 1965)

Examples of streets in the valley named after the industrialists.
(SJL and CJM)

allowance of coal per household to use free of charge, and cast slag from the smelters could be used as good quality coping stone. Throughout the valley, but particularly in Hafod, street names still reflect the association with the industrial era—Vivian Street, Grenfell Town, Jersey Street, and so on.

All Saints' Church and schoolroom near St Thomas on Kilvey Hill, built by the Grenfell family. (SJL)

The Grenfell family developed a village near St Thomas on Kilvey Hill and, like the Vivians, they built churches and established schools for their employees and their families. The Grenfells were very popular because they chose to live on the eastern side of the city amongst

Map of Swansea west, showing some buildings erected by the Vivians. (CJM)

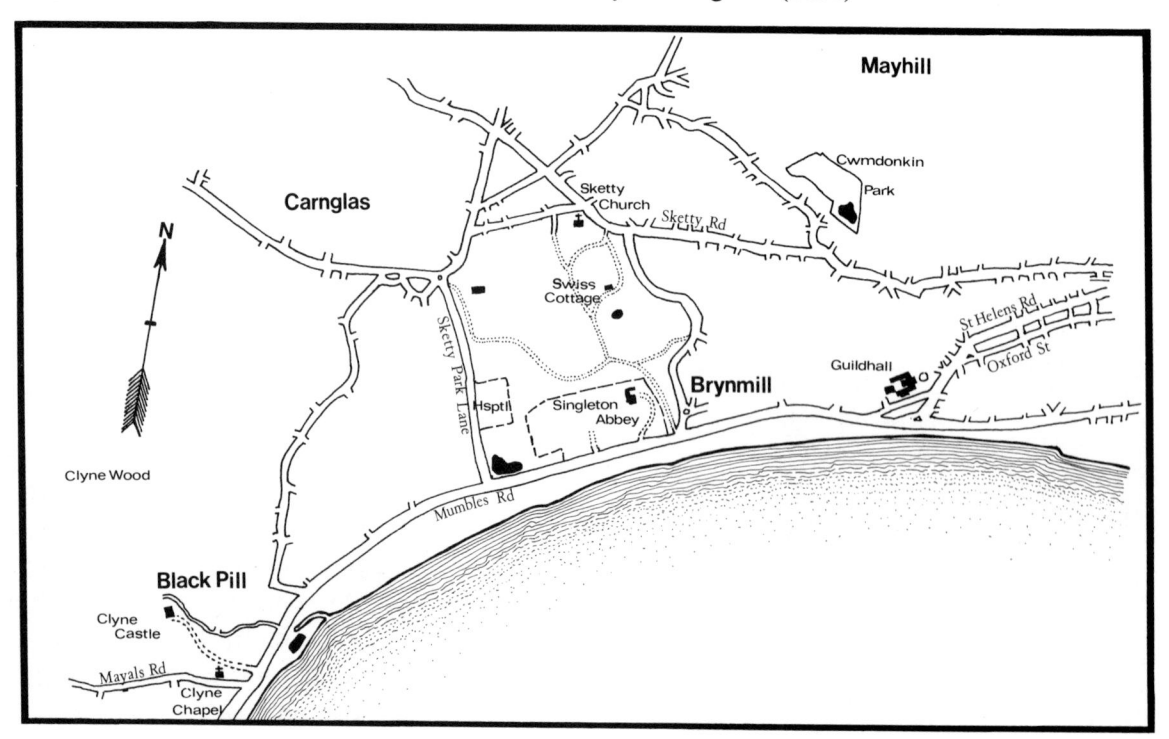

Singleton Abbey, once home of the Vivian family, which is now part of the University College, Swansea.
(SJL)

Swiss Cottage, Singleton Park, commissioned by J H Vivian.
(SJL)

their workers; the Vivians, on the other hand, preferred to get away from the industrial environment and lived in attractive surroundings on the western side of Swansea. The Vivian family used its new-found wealth to build comfortable homes such as Singleton Abbey and Woodlands Castle, Clyne; Singleton Park was established and churches were built near Sketty and Clyne. At a lecture in 1880 H H Vivian confessed that the bells for Sketty Church were made of Hafod copper which had not been paid for, but was taken by him as an honorarium for one of his inventions.

Working conditions in the smelters were harsh: the high temperatures of the furnaces, together with the smoke and dust, caused many health problems. Until 1840 the smelting works operated a system of 24-hour shifts, but in that year 12-hour shifts were introduced. In 1850 a furnaceman could earn 25 shillings (£1.25) for a week's work; women and girls were used to wheel the ores in barrows for just nine shillings (45p) a week for their work, and younger girls and boys were employed to supply beer, cider and cold water to furnacemen working at the furnaces. But despite these conditions, many hundreds of men moved to Swansea to find employment in the smelters; once they had jobs there, they had to become acclimatised quickly—within a couple of weeks—or else they were likely to die from the effects of the fumes. Contemporary photographs show how ill these workers looked, with their pale hollow faces and thin bodies, but for the sake of their families they had to continue to work in the smelters no matter how bad their health.

In spite of these conditions, however, the population of Swansea continued to rise, and by 1883 there were 71 000 people living in the city. Housing developments and sanitation facilities could not cope with such a rapid increase, and so the inevitable result was overcrowding and its associated problems, including the spread of disease. Epidemics frequently swept through communities; the under-fives were particularly vulnerable. In

1886 there was a severe epidemic of cholera; in 1869 scarlatina, which claimed 250 people; in 1870 and 1871 smallpox and typhoid fever, and in 1894–5 another epidemic of scarlatina. By 1883 the average life expectancy of an inhabitant of Swansea was only 24 years. The Port of Swansea Sanitary Authority began to make regular checks on ships returning from overseas with their cargoes of ore in an effort to prevent further outbreaks of disease, and in 1884 at least 1200 ships were inspected.

During the early twentieth century immigration continued; by 1909 the city's population had risen to over 110 000, and epidemics of measles, whooping cough and diphtheria continued to devastate the community. Immigrant workers from Ireland, Germany, Belgium, as well

Workers from the tinplate mills. (SM)

Boys from the tinplate works.
(SM)

as from other parts of Britain, settled in Swansea, resulting in a very mixed population.

The city itself prospered because of the great increase in trade; the exploitation of the valley and its people was almost complete.

When the boys went to serve in the war, girls took over in the tinplate works.
(SM)

Chapter 2

The Environment

> Landore. A spot rich in the renown of its metal and chemical works, but to the casual visitor, ugly with all the ugliness of grime, and dust, and mud, and smoke and indescribable tastes and odours.
>
> S C Ganwell, District Guide, *1880*

In the years preceding the building of the first metal smelter in the Lower Swansea Valley, the number of houses forming the village of Swansea had risen to 297. The landscape surrounding the village was particularly attractive, as can be seen from a poem written about Hafod in 1737:

> Delightful Hafod, most serene abode!
> Thou sweet retreat, fit mansion for a god!
> Dame Nature, lavish of her gifts we see,
> And Paradise again restored in thee.

and later

> Thy verdant fields, which wide extended lie,
> For ever please, for ever charm the eye:
> Thy shady groves afford a safe retreat
> From falling show'rs, and summer's scorching heat;
> Thy stately oaks to heaven aspiring rise,
> And with their utmost tops salute the skies;
> While lowlier shrubs amidst thy lawns are seen,
> All clad in liv'ries of the loveliest green:
> From every bush the feather'd tribe we hear,
> Who ravish with their working notes the ear.

Within 150 years, however, this beautiful valley was to be transformed out of all recognition by the effects of the copper, zinc, tinplate and steel industries. The first smelters did not arouse much antagonism from the farmers earning a living from the fields on the valley sides, nor did they cause any adverse comment from their workers. The population of the valley, as has been shown, rose dramatically due to the massive immigration of workers to Swansea. These people wanted work, and were prepared to suffer the hardships of the smelters to provide for their families.

Aerial Pollution

In the early nineteenth century the farmers on the valley sides started to notice a decline in the quality of their crops, and unusual illnesses in their cattle and horses. Unlike the workers in the smelters, however, the farmers did not stand by and simply accept the pollution of their land. In 1833 they contended that the devastation of the land immediately surrounding the valley was a direct result of the copper smoke from the Vivians' Hafod works, and an 'indictment was preferred against Messrs.

View over Landore in 1908. The Landore Viaduct (left) *and the Hafod tip* (right) *are both visible.* (Postcard, 1908)

Vivian, in which copper smoke was said to be injurious to animal life and destructive to vegetation'. If the farmers could prove their case and restrict the development of the copper industry, the growth and prosperity of the city would be severely curtailed, and there would be major changes in attitudes to the development of industry throughout Britain.

According to testimonies given at this 'Copper Smoke Trial' (which is fully recorded in *The Cambrian* newspaper of 16 March 1833), 'Smoke emitted from the works which existed before the great Hafod, was not

Aerial view over the Swansea Vale works, showing the distribution of solid waste material and atmospheric pollution.
(© Aerofilms Ltd, 1965)

sufficient to denude the face of the country to the same superficial extent as that to which it is now stripped.'

In 1780 there were eight smelting works in and around Swansea, and by 1830 there were fifteen, each with a large number of furnaces and a corresponding number of furnace chimneys. 'At Hafod, there were 100 furnaces; at Upper and Middle Bank, about 80; at the White Rock, about 70; each chimney delivering to the atmosphere many thousand cubic feet of copper smoke per day. The many columns of smoke gathered together to make one huge mass of dull-white clouds.'

Charles Frederick Cliffe, in *The Book of South Wales* (1848), also noted the extent of the pollution and its effects:

> On a clear day the smoke of the Swansea Valley may be seen at a distance of 40 to 50 miles and sometimes appears like a dense thunder cloud. The copper smoke is a serious nuisance . . . and has on several occasions afforded employment to the gentlemen of the long robe; tall chimneys are of little or no avail.

The composition of the copper smoke was determined whilst evidence for the trial was being collected, and the results were listed and described as follows:

> Copper and its compounds—minute even close to the works.
> Arsenic and its compounds—minute even close to the works.
> Sulphuric acid—vapour—moderate.
> Sulphurous acid—considerable.
> Sublimated Sulphur—minute.
> Hydrofluoric acid—minute.
> Fluo-silicic acid—minute.
> Coal Smoke—considerable.
>
>   The great bulk of the copper smoke is comprised of the products of the combustion of coal. The coal is more completely burnt than it is in an ordinary fire. There is, therefore, little carbonaceous soot. All the products issuing from such sources as the smelter's furnace were in the highest state of oxidation.

At the trial, farmer after farmer gave an account of the effects of the smoke upon his land and animals. The following passage is the report of part of the testimony of Morgan Morgan, who had been a farmer at Llansamlet for over 39 years.

> When he first knew that side of Llansamlet, the land was very good for corn, hay and pasture land. Cattle and horses grazed on it. It is barren now, and what grass is there, no cattle will eat it. The land is now ten times worse than it was 15 years ago. The Hafod Works had been since doubled. Heard no complaints until Mr. Vivian's works were erected. The smoke has an effect through the bones—breaks the ribs of the cattle, and produces large knobs on their legs—some lying down could not stand. Many of the farms in Llansamlet are now untenanted because there is no grass for horses or cattle. There are now no trees there. They all died. Remembers oak, ash and sycamore there. They died in less than 20 years. They died standing and were cut down afterwards. The land is now barren from the Hafod up to the Wich-tree bridge, a distance of three miles. As you travel along the road, you get the smoke into your mouth and nostrils. The smoke is very unpleasant to the smell and taste. It has the effect of making people's breath short.

Many other farmers as well as a metallurgist testified to the same effect.

The evidence for the defence, on the other hand, turned upon 'the commercial importance of the copper works, upon the national value of the articles manufactured, and upon the general comforts, prosperity, and increasing population, and importance of the town of Swansea.' After other similar testimonies came the summing up for the defence, which was reported as follows:

> Sir James Scarlett, in a speech of great eloquence and power, argued that no proof had been adduced that the diseases of the cattle, in the parish of Llansamlet, had been caused by the copper-smoke. Such effects were as likely to have followed from bad land and worse culture. The destruction of the crops might have been the consequence of

blight which, like the copper-smoke, affects the blossom. He attempted to prove that in Breconshire, far removed from the copper-smoke, the cattle became sometimes the subject of diseases, not unlike those said to be caused by the copper-smoke. Mr. Percival Johnson entered into detail to show the increase in the value of land, which had occurred around Swansea, since the erection of the copper works; how Morriston, Landore and St. John's had multiplied their houses; how the wealth of Swansea was augmenting etc.

A verdict was immediately returned in favour of the defendants (the Messrs. Vivian, the proprietors of the Hafod Copper Works).

The news of the outcome of the trial in favour of the Vivians was greeted by the working population of Swansea with jubilation, and *The Cambrian* reported the news in the following manner:

> It is with heartfelt satisfaction we announce to our readers, that the Defendant in the important case of Davies *v* Vivian, tried yesterday at the Carmarthen Assizes, and which involved almost the very existence of the Copper Trade of Swansea, OBTAINED A VERDICT. The news has diffused the greatest of joy throughout this town and neighbourhood, which has been manifested by the ringing of bells and firing of cannon throughout the day.

In the following years copper smelting continued and expanded, and Thomas Williams, in his *Report on the Copper Smoke* (1854), noted the long-term effects of the pollution:

> Testimony [of the farmers] precludes doubt as to the fact that a prodigious column of intensely concentrated smoke, pouring continuously its fumes, for a long succession of years, over a limited radius of distribution, is indeed required in order to deprive the land of its productiveness, the surface of its clothing, the soil of its latent seeds, the earth of its vital power. The farm has disappeared to give place to the manufactory; sickly cows and horses have been pushed from the scene to make room for teeming communi-

ties of prosperous human beings. Commerce has rudely usurped the seat of agriculture—the industry of thousands that of tens.

The industries continued to grow, and were lawfully able to continue to pollute the landscape. In 1861 Mr John Percy, professor of metallurgy at the Royal School of Mines, observed:

> Swansea smelters enjoy the privilege of pouring dense volumes of thick sulphurous and arsenical smoke from comparatively low chimneys into the atmosphere . . . This privilege has now in lapse of time, become an established right, which would not readily be conceded in many other parts of the Kingdom.

Early Attempts to Combat Aerial Pollution

In the years following the copper trial of 1833 the Vivians did make some efforts to alleviate the effect of the fumes on the surrounding agricultural land. One attempted solution was the building of tall chimney stacks to carry the smoke so high that it would become diluted and non-poisonous before it reached the ground. Although they did succeed in diluting the gases released to a certain extent, the high chimneys could not reduce the effects of the dust bearing the polluting particles of metal, and served only to extend the area of contamination.

A second attempt to stimulate action to bring about a reduction in the amount of fumes was an offer of a prize of £1000 'for the purpose of remunerating any person who might completely obviate the inconvenience arising from the copper smoke.' It is not certain whether the Vivians offered this prize out of their concern for the surrounding farmsteads, or whether it was stimulated by the discovery that the value of the 93 000 tonnes of sulphur lost to the atmosphere in 1848 was over £200 000 per annum, but even this inducement did not result in any effective solution. Sir Humphry Davy and Michael Faraday were both associated with experimental work in this field, but their idea to reduce the amount of smoke by passing it through a form of water filter was not successful.

A major breakthrough came in late 1865 when the Vivian works adopted the highly successful Gerstenhöffer process to reclaim the sulphur waste. Henry Hussey Vivian ordered two experimental furnaces to be built, and following their success, 43 calciners were built to treat all the ores and regulus which were sufficiently sulphurous to be dealt with by this method. The Gerstenhöffer process was based upon the principle that the heat of a furnace could be maintained by the combustion of the sulphur in the copper and zinc ores. Consequently no more atmospheric air than necessary needed to be admitted to maintain combustion, and the resulting sulphur gases would be sufficiently strong to be used in the production of sulphuric acid. The use of this process resulted in a reduction of between 38 and 47% in the amount of sulphur released into the atmosphere, although the method could only be used with certain low-grade ores; the high-grade ores continued to be smelted in the normal way and to produce the same heavy sulphurous pollution.

Attitudes to the Valley

To the traveller approaching Swansea by train over the Landore viaduct, two quotations show the differing attitudes of people to the smelters of the late nineteenth century. First, in his *Report on the Copper Smoke* (1854), Thomas Williams described the valley thus: 'The scene fills the eye of the spectator standing on the Landore viaduct, on a tranquil summer's night fall, when the beams of the down-going sun are fringing with fire and gold the revolving clouds which veil the blackness of the slag mountains, and multiply by shadowing the forests of chimneys, peering above the slattered roofs of the vast "sheds" dubiously looming in the darkness of the valley, is really grand, if not awe-striking; sublime, if not terrific.'

Secondly, the winner of a railway carriage eisteddfod held on the return journey from Newport in 1897 gained this impression:

> It came to pass in days of yore
> The Devil chanced upon Landore
> Quoth he, by all this fume and stink,
> I can't be far from home, I think.

View south over the Lower Swansea Valley towards the city centre. It is possible to see large areas of unused land in the foreground and a number of waste tips across the middle of the photograph.
(Victor Hopkins, 1965)

The bare eroded soils of the north-eastern part of the Project area.
(LSVP 1966)

Looking back, it can be seen how quickly the effects of the copper waste destroyed most of the vegetation of the Swansea Valley. As the shelter and food provided by the trees and wild flowers disappeared, so the habitats of many birds and animals were also destroyed. Together with the high rainfall of the valley (about 1200 mm per annum) and the absence of plant roots to bind the soil together, the topsoil was gradually eroded away until the area became almost completely devoid of life. By the time the industries started to close down in the 1920s and 1930s, the whole area was yellow and sterile. In addition, thousands of tons of solid tip waste, another unwanted by-product of the copper, zinc, tinplate and steel works had been dumped into the valley, and this only added to the terrible pollution of the landscape.

Solid Tip Waste

The extraction of a metal from its ore involves the elimination of impurities which are drawn from the furnaces together with the burnt coal as slag. The slag from copper smelters contained silica, cupric oxide, ferrous oxide, aluminium and a small amount of copper.

From the earliest days of smelting in the valley this waste material was simply dumped outside the back doors of the works. With the smaller works the tipping posed a relatively minor problem, but as the smelters grew in size and capacity, so the problems of such large amounts of slag increased. The larger smelters never really came to terms with the problem of disposal; they simply bought large areas of land adjoining the works where this material was dumped year after year.

The White Rock tip in Foxhole was formed as two hundred years of dumping destroyed part of Kilvey Hill and produced an enormous tip with over 300 000 tonnes of waste which was visible from any elevated part of the valley. The Vivians' Hafod tip, reputed to be one of the

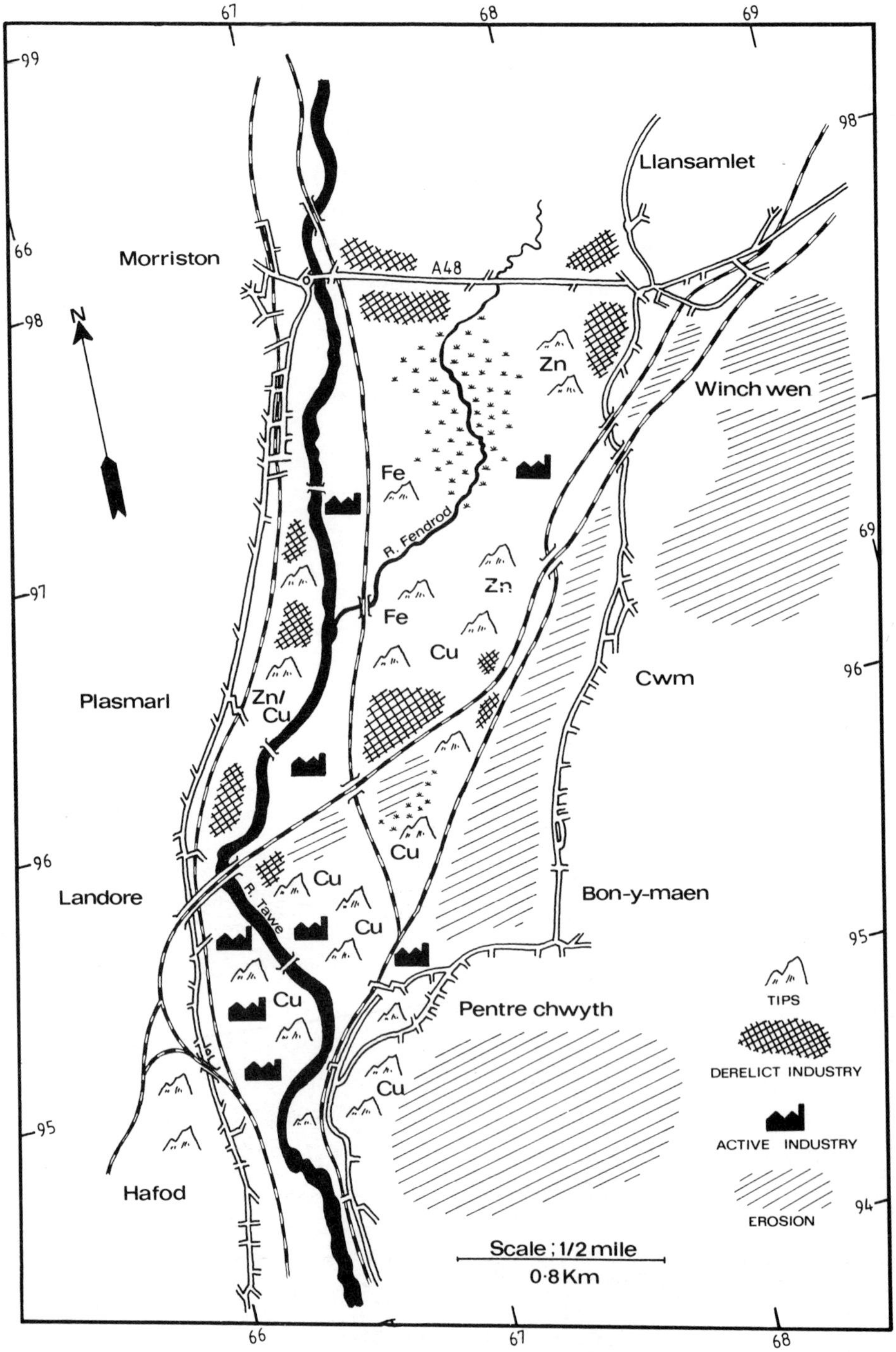

Llansamlet
Morriston
A48
Winch wen
Zn
Fe
R. Fendrod
Zn
Fe
Cu
Cwm
Plasmarl
Zn/
Cu
Cu
Cu
Landore
R. Tawe
Cu
Bon-y-maen
Cu
Pentre chwyth
Cu
Hafod
TIPS
DERELICT INDUSTRY
ACTIVE INDUSTRY
EROSION
Scale ; 1/2 mile
0·8 Km
67
68
69
99
66
98
97
96
95
94

Map of the Lower Swansea Valley in 1960 showing the distribution of tips, dereliction and erosion.
(CJM)

View over the River Tawe towards the Hafod and Morfa works. The tips from both smelters can be seen.
(LSVP 1963)

highest copper slag tips in Wales, covered a five-hectare site alongside the houses built specially for the workers.

Throughout the valley, waste tips of various hues were produced by the industries: orange from the copper smelters; black from the lead and zinc works; and later on, lighter-coloured waste from the steel works. The slag gradually poisoned the ground, the immense weight of the tips compacted the underlying soil, and wind and rain eroded the material into grotesque shapes. It was estimated that the total amount of waste in the valley left after the smelters closed down was seven million tonnes.

Again attitudes to this slag left in the valley differed. Two passages taken from Thomas Williams' *Report on the Copper Smoke* (1854) illustrate this ambivalence. First,

... a smiling valley has at length, indeed been transformed into a desert scene of ashes and lifeless gravel tracts. The mountains of scoriae, consisting chiefly of the silicates of iron, coal, ashes and other refuse products which spread over many superficial areas in the vicinity of the works, are utterly infertile. Black and uninviting to the eye, they are capable of lodging no single form of life. The hardy moss clings to, and even thrives upon the soilless rock. It dies on the copper-slag bank.

A copper waste tip in the centre of the valley with sparse grass cover. (SJL 1979)

On the other hand, in the same report, Williams attempted to find some virtue in the mounds of slag:

Passive, harmless mountains of silicaceous and metallic ashes, they do good by negation. They suppress malaria. They have obliterated a tidal morass. Even to the eye of cultivated taste, a copper-slag bank, with all its charmlessness, is preferable to an aguish marsh. To the philanthropist, thoughts like these are consolatory.

The tips did indeed have some small value to the local residents, who would sift through the waste searching

The White Rock copper tip. (LSVP 1963)

View over the Morfa copper tip towards Landore and the ruins of Morris Castle.
(LSVP 1965)

for half-burnt coals which could be used again in the home, and for lumps of cast slag for hardcore and for use as coping stone. These practices were common wherever smelting and mining operations took place in Britain.

Many other heavily industrialised regions of Britain were also experiencing the effects of pollution. In 1937, in *The Road to Wigan Pier,* George Orwell wrote of the north of England: 'You begin to encounter the real ugliness of industrialism—an ugliness so frightful and so arresting that you are obliged, as it were, to come to terms with it.' Of Wigan itself he commented: 'All around was the lunar landscape of slag heaps, and to the north, through the passes, as it were, between the mountains of slag, you could see the factory chimneys sending out their plumes of smoke . . . It seemed a world from which vegetation had been banished; nothing existed except smoke, shale, ice, mud, ashes and foul water.'

Over the course of time, another process was taking place underneath the mountains of copper waste. The high rainfall of the valley not only eroded the tips into ugly shapes, but also leached some of the toxic copper and zinc salts into the underlying soil. The resulting

Scene of desolation in the Swansea Valley—mountains of copper waste.
(LSVP 1963)

contamination inevitably posed many very serious problems for the botanists in their attempts to revegetate the valley when the tips were eventually removed. The various methods used to try to resolve these difficulties are discussed in Chapter 4.

Pollution of the River Tawe

We have seen how the air and land in the Lower Swansea Valley were seriously affected by the waste products from the various industrial processes, but in addition to all this, the once clear River Tawe, which had been well stocked with fish, also became polluted.

The river had been vital as a transport medium to keep the smelters supplied with ores; it provided water for use in many industrial processes; but it also provided an easy means of disposal of effluent. Liquid waste from the alkali and copper works, sulphuric acid and iron

The River Tawe. On the banks, derelict buildings and railway sidings, and in the river itself, tip material and the decaying hull of a ship.
(LSVP 1963)

sulphate from the tinplate works, together with ashes from the iron works and cinders from the tips were poured into it, and all combined to destroy the life of the river.

Thomas Williams' observations of the River Tawe in 1854 illustrate the extent to which it had become polluted: 'A small dirty, muddy river, capriciously appearing and disappearing, as it may plunge into or

Ruins of the White Rock works on the banks of the River Tawe.
(LSVP 1963)

emerge from the rolling masses of smoke creeping along its "slag" embanked channel.' And later, '. . . The low places along the river banks would be changed into exhalent pools of rank vegetation. The stagnant waters would become slimey, and fevers would decimate a hapless population. This is no picture dictated by poetry. It comes of observation.'

View over Morriston showing the derelict works, tips and the effects of flooding in the valley.
(LSVP 1964)

Dereliction at the White Rock works.
(LSVP 1963)

By the time the industries closed during the 1930s the pollution of the river was complete and all plant and animal life there had been destroyed. The river which had given the valley its life now ran like an open sewer between the mountains of poisonous slag and the derelict industrial buildings.

The derelict laboratory of the Siemens works.
(SM)

Dereliction

> ... in the industrial areas one always feels that the smoke and filth must go on forever and that no part of the earth's surface can escape them. In a crowded, dirty little country like ours one takes defilement almost for granted. Slagheaps and chimneys seem more normal, probable landscapes than grass and trees ...
> *George Orwell,* The Road to Wigan Pier, *1937*

In the years immediately following World War I there was a short-lived revival of the tinplate industry, but the majority of other works began to close down. The only zinc smelter in operation in the 1920s was the Swansea Vale works (which eventually closed in 1974), and within the next few years all the remaining copper smelters ceased production. The two major works still operating in the valley in 1940 were Yorkshire Imperial Metals (a

The derelict White Rock works and tip.
(LSVP 1962)

Derelict zinc smelters and tip waste as seen from the main railway line into Swansea.
(LSVP 1962)

joint company of Yorkshire Metals and ICI, occupying the site of the Hafod works), and the British Steel Corporation's Landore works (the old Landore steel works owned by Richard Thomas and Baldwins Ltd), but both of these were closed down in 1980, thus severing all remaining industrial links with the past.

A derelict zinc smelter.
(LSVP 1963)

Course of Smith's Canal in 1978. (SJL)

Bollard from the shipping era now stands in land used as a car park, and the Weaver building, now derelict and dangerous, towers over the filled-in dock. (SJL 1979)

The smelters were abandoned; in many cases their owners could not be traced, and left behind problems caused by their decaying buildings and waste tips. As the years passed by, the old buildings deteriorated further in the wind and rain, and also at the hands of vandals. The ugly mounds of tip material were made even more grotesque by the indiscriminate removal of bits and pieces for hardcore.

In the 1920s a great deal of coal was still being exported from the port of Swansea, but very little was in fact mined in the locality—the last pit in the Lower Swansea Valley was closed in 1931. As the industries declined, so the canals also fell into dereliction. Although they were not filled in until the 1960s, no barge used either Smith's Canal or the Swansea Canal after the 1930s, and an air of apathy gradually settled over the valley. In 1969 John Barr, in his book *Derelict Britain*, noted that 'Nowhere in derelict Britain is there a more dismaying example of man creating wealth while impoverishing his environment than in the Lower Swansea Valley . . . the Lower Valley has been often called in the past the most concentrated and uninterrupted area of industrial dereliction in Britain.'

The last link with the copper industry of the 1800s was finally broken in 1980 with the closure of the Hafod works. Many of the chimney stacks have now gone and the Swansea Canal (foreground) has been filled in (compare with the drawing of the same scene on page 9).
(SJL 1979)

Early Reclamation Plans

It was back in 1912 that the problems of the Lower Swansea Valley were first officially recognised, when George Bell, the borough surveyor, put forward a major scheme entitled *Floreat Swansea* to the city council, for the redevelopment of the area.

Bell had observed the problems caused by the floodwater in the valley and its effects, not only on the landscape, but also on the poor-quality footpaths and roads to the works (flooding of the valley became a major problem after the tipping of waste material on the valley floor seriously disrupted the natural drainage channels). Some routes were completely impassable after flooding. Bell suggested that a new road network should be built to link the east and west banks of the river to provide better communications, and which would also open up the already derelict areas for redevelopment.

His report recommended the use of waste material lying around the valley to raise the level of the land to prevent flooding, and even to use it for road ballast. He envisaged new houses and industries laid out on 'garden suburb' lines. New tramways could link existing communities and would give people 'full benefit of all the institutions of Swansea for business, education and the recreation of pleasures of life.'

Unfortunately, Bell's ideas were far ahead of his time, and his imaginative proposals were ignored, although some of his ideas and principles were eventually incorporated into the later plans for the Lower Swansea Valley.

After the closure of most of the works in the 1930s the valley had become a barren, silent, industrial desert. The amount of wasteland had increased steadily and no attempts were made to revive or reclaim old sites by introducing new industry. Until the widespread problem of unemployment arose in that decade, there had been no strong national policy with respect to derelict land. However, the situation demanded a government investigation into the nation's depressed areas, as a result of which the Special Areas (Development and Improvement) Act of 1934 created a fund of £2 million to help relieve the social and economic problems of these areas.

Although just under one quarter of Swansea's adult population was unemployed in 1931, the area was reasonably fortunate when compared with the rest of Wales. The number of unemployed people in Wales as a whole reached 36.5% in 1932, but in South Wales and Monmouth the figure was an incredible 41%. The coalmines of the South Wales valleys were closing down, but there was still a considerable demand for the anthracite coal being mined around Swansea and exported from the port. According to the terms of the Act, which excluded areas of relatively low unemployment, Swansea did not qualify for grant aid and no reclamation work could be started. More satisfactory legislation was postponed following the outbreak of World War II in 1939.

In 1940 a report by the Royal Commission on the

Distribution of Industrial Population recommended more government involvement in the development of the industrial growth of the 'special areas', and influenced government policy which resulted in the Distribution of Industry Act of 1945. The old 'special areas' were grouped into larger regions and Swansea, as well as Newport and Cardiff, were now included. For the first time the reclamation of derelict land became the subject of grant aid. This Act appeared to provide old industrial regions with the stimulus of special powers and grant aid to clear up their landscape and develop new industries.

In Swansea, the council's borough engineer in 1943 recommended that if government aid were to be made available, then the city council should acquire and clear the derelict sites in the valley to provide sites for new industry. Secondly, he recommended that the council should consider the development of a light industrial estate. In June 1945 the Distribution of Industry Act gave the Board of Trade all the powers necessary for the development and management of industrial estates. Swansea city council therefore decided to develop a light industrial estate at Fforestfach, to the north-west of the city, in November 1945, and the responsibility for this was eventually taken over by the Board of Trade.

The council then concentrated on the acquisition and clearance of the sites in the valley. The Distribution of Industry Act empowered the Board of Trade to acquire derelict land in development areas and to carry out such work as was necessary to reclaim the land so that it could be used by the community. Once again, however, as with the Special Areas Act of 1934, the relief of unemployment was the major consideration. With the establishment of the industrial estate of Fforestfach and the relatively low unemployment in the county borough, the government decided that further funds for the preparation of more industrial sites in the area were unjustified, and grant aid was refused.

In 1945 the city council had its own share of problems

and there were other projects to deal with before it could begin any reclamation work using its own funds. First of all, it was committed to building the industrial estate at Fforestfach; secondly, it was already carrying out a major programme of modernising the docks; and thirdly, and most important, was the rebuilding of the city centre which had been heavily bombed in 1941—most of the city had been completely flattened, and its redevelopment was consequently put at the top of the agenda. As a result, the reclamation programme for the valley gradually fell down the list of priorities.

The centre of Swansea after the bombing of 1941.
(SM)

During the 1950s the government was forced to cut back on its capital expenditure because of Britain's acute economic problems, and thus grants for the reclamation of derelict land were severely limited. In 1956, as their financial restrictions were gradually eased, local authorities were again asked to submit their plans for reclamation. This time, however, the very word 'derelict' was to be a stumbling block to Swansea's case. Most of the land in the valley was privately owned, and by declaring the land derelict the landowners would in fact be stating that the land was worthless and of no value to them. Naturally, no landowner in the valley was prepared to make such a declaration and consequently none of the land became eligible for grant aid.

The Local Employment Act of 1960 replaced the 1945 Distribution of Industry Act and widened the definition of the word 'derelict' to include neglected and unsightly land. Once again aid was based on the criterion of whether 'a high rate of unemployment exists or is to be expected . . . and is likely to persist'—Swansea's continuing low level of unemployment compared with the rest of Wales again prevented Britain's most concentrated area of industrial dereliction from qualifying for grant aid.

By this time, it had become obvious that no grant aid was likely to be allocated by central government. Locally, there were a number of reasons why there was no incentive to clear the derelict landscape by levying a special rate to provide the necessary finance. In the first place there was no real need for more industrial sites; secondly there was no lack of land for development in the city; and thirdly there persisted in Swansea a deep resentment against those who had destroyed the valley and had then withdrawn to other parts of the country. This was quite a turn-around from the residents' feelings towards the industrialists when their livelihoods had been threatened at the Copper-Smoke Trial of 1833. Finally, the complexity of the area—the tips, derelict buildings, multiple ownership of land, and the physical fragmentation of the valley by canals, the river, and railways—led to the belief that the solution to such an enormous problem was beyond the resources of the county borough. Apathy set in, and when the last tinplate works closed in 1961 it seemed likely that the valley would never recover from its industrial despoliation.

Chapter 3

The Lower Swansea Valley Project

... to establish the factors which inhibit the social and economic use of land in the Lower Swansea Valley, and to suggest ways in which the area should be used in the future.
K J Hilton, Lower Swansea Valley Project Report, 1967

The Lower Swansea Valley Project is like an explorer's torch in the gloom of a cave. It reveals the mess, but it also gives a glimpse of the possibilities . . . it might well encourage the inhabitants of other parts of the country to feel the need to do the same.
Prince Philip, Lower Swansea Valley Project Report, 1967

The Initiative

The approach to Swansea by train, through the valley and over the Landore viaduct, has moved many writers and poets to describe their impressions in the most vivid terms, as shown in Chapter 2. Whilst land often remains derelict for many years because it is out of the way or in a part of a town or city that no one visits, this could not be further from the truth in the case of Swansea. Although no major roads cross the lower valley, the main London to South Wales railway line cuts through the heart of this desolate landscape. It is said that some travellers visiting Swansea for the first time saw this tragic scene and returned to London without leaving the station; others who knew Swansea were happy to shut their eyes as the train crossed the valley and would hurry off to their comfortable homes on the west of the city. It appeared, certainly, that most people in Swansea gladly ignored the valley, and simply wished it out of existence.

There was one person, however, who appreciated the problems but refused to ignore them, Robin Huws Jones, then the director of courses in social administration at the University College of Swansea. He was convinced that a new and independent initiative was required if the problems of this derelict and forgotten landscape were to be solved. He suggested that the base for this fact-finding study should be the University College, but carried out in collaboration with the Ministry of Housing and Local Government, Swansea Borough Council and all other interested groups. The initiative of this one man was to be responsible for breaking the apathy which had existed for so long, and for establishing a comprehensive survey which was to be the first step toward the reclamation of the Lower Swansea Valley.

The Project

Robin Huws Jones' initial enquiries in the late 1950s drew a favourable response and the then Principal of the University College, John Parry, saw the importance of the study not only to Swansea, but also to the college. Apart from being part of solving a long-term problem, the college would be able to carry out a unique inter-disciplinary study of the area. The college was favoured as the base for the investigation by the borough council, the Welsh Office and industry, as the problem would then be taken out of local politics. The college itself was based in Singleton Park around the Abbey which for many years had been the family home of the Vivians.

An informal working group was established in 1960 under the leadership of the principal. Heads of college departments who gave their support were joined by representatives of the borough council, the Welsh Office and industry. Their first problem was to secure enough money to carry out a detailed enquiry. It was estimated that £48 750 would be required to fund a four-year study, and half this amount was eventually met by grants from the then Department of Scientific and Industrial Research, the Welsh Office, the Ministry of Housing and Local Government, and Swansea Borough Council. The Trustees of the Nuffield Foundation then also agreed to contribute an initial sum of £22 500 towards the enquiry, and the University College provided the office, laboratory and technical facilities required, which was a substantial donation in kind. With its finance assured, the working group became a committee, and in August 1961 Kenneth J Hilton was appointed as the executive director of the Lower Swansea Valley Project.

The ultimate aim of the Project was the complete reclamation of the valley, and although they did not have the funds to complete this work, it was hoped that the academic study, together with a limited amount of practical action, might serve to break the long-standing apathy towards the area, and at least provide a full and complete source of information for the city council planners. The Project's terms of reference were 'to establish the factors which inhibit the social and economic use of land in the Lower Swansea Valley and to suggest ways in which the area should be used in the future.' By investigating the physical, social and economic problems of the valley, it was hoped to be able to understand the reasons which had inhibited development in the past, and to provide information and recommendations for its future development.

The organisation and coordination of the four-year study were carried out by the director of the project and through the main committee, the steering committee, the academic sub-committee, the land-use sub-committee and the editorial sub-committee. Six college departments were involved with the study, and research teams were specially appointed to work on particular aspects of the programme, working as members of the appropriate college departments.

Studies of the Project

The studies carried out by the research teams can be divided into two main groups: (*a*) physical studies (biology, soil mechanics, hydrology, geology and contour mapping); and (*b*) socio-economic studies (employment, housing, recreation and land use). The study area was defined as a roughly inverted triangle, three miles (4.8 km) long and a mile (1.6 km) wide at its northern boundary, covering 475 hectares. The villages around the boundary of the study area included Hafod, Plasmarl, Morriston, Llansamlet, Winch Wen, Bonymaen, Cwm, Pentrechwyth and St Thomas. As mentioned in Chapter 2, the valley was badly fragmented by being in the ownership of 44 different landowners, and also by the river, canals and railway lines.

(*a*) *Physical studies*

Geology, soil mechanics and foundation engineering. This study described the geology of the Project area and its immediate surrounds, and investigated the problems associated with the large areas of soft and compressible natural deposits; the influence of industrial waste as fill material, and the dangers from the abandoned coal workings to future building projects.

A detailed geological map at a scale of 1:5000 was drawn up, and this enabled the quantities of fill material to be estimated when compared with the geophysical traverses across the valley floor which revealed the approximate shape of the buried valley. It was observed that the dumping of tip waste had in fact served to reclaim large areas of marshy land by improving the foundation conditions. There was no evidence that any existing buildings had suffered as a direct result of subsidence due to mining.

The study also provided an accurate account of the volume and cover by tip material and observed that 50% of the Project area was covered by non-ferrous slag, iron and steel slag, coal tips, factory waste, building rubble and recent urban tipping.

Revegetation techniques. The Project's botanists identified the three principal types of derelict land in the valley as: 'infertile and largely non-toxic clay loams badly eroded by former smelter smoke pollution; tips of relatively innocuous waste material (coal slag, foundry sand, domestic refuse); and tips of waste poisonous to plant growth and derived from the smelting of copper and zinc ores.' This study set itself the problem of developing 'techniques for the inexpensive establishment of vegetation giving an acceptable cover and requiring minimal maintenance.'

The botanists established large-scale trial plots on copper, zinc and steel waste and carried out experiments with cheap sources of fertiliser and normal grass seed species, as well as smaller-scale experiments on waste brought into the University's greenhouses. The revegetation of copper and zinc tips proved to be a difficult problem, but two methods proved encouraging. Firstly, by incorporating lime, a general fertiliser and a cheaply available form of organic matter such as sewage sludge into the tip material, it was found that certain plants (common bent grass, gorse, and the genus *Buddleia*) could be established. Secondly, certain strains of grasses (especially common bent grass and creeping bent grass) have evolved a degree of tolerance to the metallic poisons in the tips. These grasses were further encouraged by the addition of a general fertiliser and organic material.

Trees were planted directly into tip waste and into the eroded soils, but unfortunately it has not been possible

Geophysical studies in the shadow of the Swansea Vale works.
(LSVP 1963)

to show the successful establishment of any tree species in tip material, although forest plantations have been established on the poor soils using normal forestry techniques.

Since the botanical work of the Project ended, a great deal of further research has been carried out at university botany departments throughout the country on the problems provided by the study. Investigations into tolerant species of plants and their methods of tolerance were current research topics in 1980. In the valley the continuing growth of the forest trees provides a constant source of study into the problems of trees growing in highly contaminated soils.

The River Tawe. The River Tawe has a rapid alternation of high and low flows, described as 'flashy'. Flooding caused by high rainfall, tidal obstructions (weirs, bridges and culverts) and the filling of flood storage space by tip waste has caused serious problems in the Project area. Pollution of the river had been grave but the position gradually improved. The hydrologists recommended that a small tributary of the Tawe, the Nant-y-Fendrod, should be culverted, and the valley protected from flooding by a levée system. They believed that the river should be regarded as an amenity, and that the erection of a barrage to hold the water level at high tide would encourage the development of a riverside park eventually to provide for water sports such as rowing and canoeing.

Highway and transportation planning. This study formed part of the work carried out by the civil engineering department of the University College. It noted that classified traffic routes bordered the valley on three sides, but the floor itself contained only short, badly surfaced and often unconnected works access roads. The research team was faced with the problem of two disused canals, a criss-crossing of poor roads, a river and a series of railway lines, and so they carried out a number of

Early botanical work on trial plots with various grass species and fertilisers.
(LSVP 1963)

surveys to study regional and local road traffic movements. From their findings, the study groups recommended that the Swansea Canal should be filled in and that a completely new road network should be built on the valley floor. It was also suggested that a new system of pedestrian footpaths connecting surrounding residential areas with the new amenity areas in the valley was required.

(b) *Socio-economic studies*

The people, the environment and housing. The Project's sociologists discovered that according to the parameters of age, sex, marital status, occupation, status grading, educational attainment, income, receipt of free school meals, of children taken into care, and of juvenile delinquency, the population of the Lower Swansea Valley was a normal cross section of the County Borough of Swansea. However, on the basis of such factors as public buildings (e.g. schools), the standard of housing and the amount of open space, this normal population was living in a substandard environment. Compared with Swansea West, having almost three hectares of recreation space per 1000 population, and the national recommended figure of 2.4 hectares, the provision of less than one hectare per 1000 in Swansea East was sadly deficient. In addition, even this small amount of open space was found to be of a lower standard than that in Swansea West.

Despite the decline of the valley's industry, the population has remained roughly static, even showing a slight increase in recent years. The study group noted that there was a need to build new houses in the borough and to improve many of the existing ones. They believed that large areas of derelict industrial land could be developed for residential and amenity use, rather than for new industrial sites.

It was this study which most favoured the idea of establishing parkland in the valley to link in with the

Early Project work—a weather station in the valley. (LSVP 1963)

forest plantations which would be planted by the botanists. The development of a riverside walk, the formal use of certain tips as adventure playgrounds, and bicycle tracks, were all new ideas for the use of this hitherto abandoned landscape. Finally, the group recommended that playing fields and a sports complex should be considered as a part of a regional plan for the improvement of sports and recreational facilities.

It was also noted that any further industrial development should not involve any factories which would be likely to create severe air pollution because of the large increase in the amount of land designated for residential use.

Prospects for industrial land use. This study group considered the question of the use of part of the valley for industrial sites. Since the Lower Swansea Valley is centrally located in a region of almost half a million people, the group concluded that with careful planning and development, certain sites could be very valuable to industry, but advised that any such developments should be made within a full and comprehensive plan for the revitalisation of industry in South Wales as a whole.

Visual improvements. The Project's recommendations for improving the visual appearance of the valley were as follows: 'The visual improvement of derelict land is an essential first step towards its eventual redevelopment for industrial, residential and amenity use . . . the continuing scale of dumping of household rubbish, builders' waste and abandoned cars must be stopped. Schools must be encouraged to take part in efforts to upgrade the environment. In the Swansea Valley, vandalism has been reduced as a result of schools' participation in tree planting and conservation. This kind of activity needs to be continued on a more permanent basis.' To encourage this community involvement a

Conservator was appointed, and his role is discussed in the next chapter.

Use of derelict land. This study suggested that the redevelopment of the valley floor should be an integrated programme of housing (46.5 hectares); industrial and commercial use (165 hectares), and amenity (48.5 hectares), although it was later decided that the area designated for industry could be reduced by 32 hectares to provide room for a sports complex. The new industry would have to be non-polluting, and any new housing or renovated old residential areas should be part of a comprehensive plan. The Report recommended that the parts of the valley not scheduled for reclamation in the immediate future should be temporarily landscaped to provide a more pleasant prospect during the interim period, and that all of the newly planted areas should be maintained.

Economic assessment of land use proposals. An economic assessment of the land use proposals was made, which stated that the valley should be treated as a single project '. . . with the land to be acquired and developed by a single body. Whilst the total reclamation of the valley is unlikely to be completed for some years, in the long term the authority developing the whole area is likely to get a good return on its investment. There is a strong case for central government grants towards the cost of reclamation.'

The Lower Swansea Valley Project Report

All studies were completed during 1966, the Project Report was edited by the director, K J Hilton, and was published in 1967. The Report was based on the twelve study reports outlined above, the titles of which are given below.

(i) *Human Ecology of the Lower Swansea Valley* (M Stacey)
(ii) *Report on Transportation and Physical Planning* (R D Worrall)
(iii) *Report on the Hydrology of the Lower Swansea Valley* (D C Ledger)
(iv) *Report on the Geology of the Lower Swansea Valley* (W F G Cardy)
(v) *The Soil Mechanics and Foundation Engineering Survey of the Lower Swansea Valley* (H G Clapham, H E Evans and F E Weare)
(vi) *The Prospects for Industrial Use of the Lower Swansea Valley—A Study of Land Use in a Regional Context* (S H Spence)
(vii) *Lower Swansea Valley: Housing Report* (M Stacey)
(viii) *Lower Swansea Valley: Open Space Report* (M Stacey)
(ix) *Plant Ecology of the Lower Swansea Valley*
    (*a*) *Vegetation Trials* (R L Gadgil)
    (*b*) *Large-scale Grass Trials* (L J Hooper and R Garrett Jones)
    (*c*) *Soils* (E M Bridges)
(x) *Soil Biology of the Lower Swansea Valley* (P D Gadgil)
(xi) *Afforestation of the Lower Swansea Valley* (B R Salter)
(xii) *Tips and Tip Working in the Lower Swansea Valley* (G Holt)
Estimation of Quantities of Materials in the Northern Part of the Lower Swansea Valley Project Area (H E Evans)
Feasibility of Creating an Artificial Lake in the Lower Swansea Valley (R E Davies)

Complete sets of these reports were deposited at the University College of Swansea, the Guildhall, Swansea, the National Library of Wales, Aberystwyth, and at the British Museum in London.

When the Project Report was published in May 1967 it was welcomed by the local press. Its conclusions and recommendations were clear and concise: The vast derelict area of the Lower Swansea Valley should in future be regarded as a valuable new resource, and should be redeveloped along the lines put forward in this blueprint. The first hurdle to be overcome, as detailed in the Report, was for a single authority to take on the job, acquire the land, and produce a full-scale plan for its reclamation. The University College did not have the funds to carry this out, Hilton stated in the Report, and continued, 'We, the Project Committee and its sponsors, who have seen the proposals develop through the last five years, are concerned that these pages should not be its epitaph; and that others will carry forward where we have left off, not only in the Lower Swansea Valley, but wherever derelict land and a poor urban environment are found.' Swansea's town clerk in May 1967 announced the council's determination 'to get on with it'.

Planning in the Valley

It was back in 1912 that George Bell first put forward his ideas for the redevelopment of the Lower Swansea Valley in *Floreat Swansea,* but later, in 1944, the *Picture Post* proudly proclaimed 'We plan a valley in South Wales.' The article vividly described the valley's problems, and then pictured it 'as it might one day look. It is a dream—but a dream based on reality. It depends on the government which must make the first decision.' The plan was wildly ambitious, and made no economic assessment of the cost, so that the whole idea was taken badly by the local council of the day.

The County Borough Development Plan of 1960, proposed under the terms of the 1945 Act, stated that 'The Tawe Valley is to remain for industrial use, mainly for heavy industry, including the tipping of industrial waste to controlled levels for future use.' Following the Aberfan disaster of 1966 and the publication of the Lower Swansea Valley Project Report in 1967, local and national concern with respect to derelict land and tipping was aroused. Under the Industrial Development Act (1966) and the Local Government Act, grants were made available to local authorities for large-scale land improvements, and the County Borough of Swansea initiated a major programme of land reclamation.

Arising from these new developments, in 1968 the county borough prepared a draft development plan which for the first time suggested uses for land in the valley other than just for industrial expansion, and also that a new road network should be built. Land to the north of the main London to Swansea railway line should be allocated for industrial use, and land to the south (including Kilvey Hill) should be developed as 'open space' areas, including woodlands and other recreational facilities. The improvement of the River Tawe was another major recommendation of the 1968 plan.

Between 1968 and 1974 a number of reclamation schemes were carried out, and following local government reorganisation in 1974, Swansea City Council published an informal local plan for the valley called the Interim Planning Statement. This plan extended the boundaries of the Project area to include all the land from the mouth of the Tawe in the south to the M4 motorway in the north. The overall plan for the valley now had six interrelated detailed schemes, including a barrage of the river, and these are outlined in Chapter 5.

Land Acquisition

The city council gradually began to buy plots of land in the valley and to put into operation various reclamation projects. In 1964 the council owned 1.38 hectares, but by December 1978 it had acquired almost 300 hectares. By 1981 the council either owned or was negotiating for the remaining land requiring development in the valley, and was thus able to put forward other detailed proposals for the future. Up to 1979 the city council had received over a quarter of a million pounds in grant aid from the Welsh Office for land acquisition, and since then financial assistance for this purpose has been included as part of the council's reclamation schemes for approval by the Welsh Development Agency; by March 1979 the council had received £580 000 from the Agency. Generally, land acquisition has been by voluntary negotiation, but statutory powers of compulsory purchase have been used to acquire some small plots of land.

The Project Report was an essential first step towards making the city of Swansea face up to the problems of dereliction in the valley, and consequently bring about a start to its reclamation. Throughout Britain attitudes towards the environment were changing in the late 1960s and early 1970s, grant aid was becoming available, and a greater effort was being made to rid many parts of the country of their tips and to improve their neglected landscapes.

Chapter 4

Reclamation

It is so easy to justify dereliction and pollution in the name of progress; it is so simple to lay ruthless hands on the unresisting countryside in the name of national interest. It is nothing like such a simple matter to clear up the mess afterwards.
Prince Philip, Lower Swansea Valley Project Report, 1967

The Lower Swansea Valley Project was determined that its academic survey and wide-ranging investigations would bring about as much effective visual improvement as its limited finances and resources would allow. The derelict industrial buildings, the unsightly tips and barren ground were the main targets for reclamation. Almost as soon as the Project came into being in 1961, volunteers and other groups were encouraged to become involved practically in helping to rid the valley of its derelict appearance.

Early Voluntary Work

Swansea was the headquarters of the 53rd Divisional Engineers Territorial Army regiment. Whilst the Project did not have the mechanical plant to demolish the derelict buildings in the valley, the Territorial Army did, and were able to use the derelict sites as valuable training grounds for handling their bulldozers, etc, and for demolition practice. Through cooperation between the

Removal of part of the White Rock tip by the local Territorial Army.
(LSVP 1965)

Unit's Commanding Officer, the Project and the local authority, a great contribution was made to the clearance efforts. The regiment started work in February 1962; they dismantled six hectares of derelict stone and brick buildings running alongside the London to Swansea railway line, and demolished the old Llansamlet copper and arsenic works with a series of small explosive charges. Bulldozing and flattening of the land was carried out in February and March 1962, during which time weekend camps were organised for the Regular and Territorial Army Engineering Units. During August 1962, derelict buildings on the three-hectare site of the Dillwyn works were demolished. In both these cases it was hoped that other landowners would follow up the operation using their own resources to complete the clearance.

In May 1963 a troop of the 48th Squadron Royal Engineers started the demolition of the former Morriston spelter works, but unfortunately the explosives also

Removal of a derelict smelter by the Territorial Army.
(LSVP 1964)

Demolition of unsafe chimneys by the Army.
(LSVP)

caused damage to windows of surrounding houses, and so this project had to be discontinued.

School children were encouraged to take part in the tree-planting work, and their involvement is described later in this chapter. In addition, the International Voluntary Service held summer work camps in the valley with British and European students working on stone picking and tree planting.

Site of the Morfa tip in 1979 prepared for new, light industries. Plasmarl and the ruins of Morris Castle can be seen in the background.
(SJL 1978)

The Photographic Record

Throughout the valley, before any demolition work was carried out, a full photographic record was made of the old buildings and their sites. Today the photographs form a valuable record for young people who have no recollection of how desolate the area was. It is so easy to forget exactly how places once looked, and the photographs are a valuable aid to illustrate to each new group of school children just how much the environment around their homes has changed in such a short time. The photographic record alone shows how depressing the valley was, with its tips, derelict buildings and accumulation of household rubbish and fly-tipped waste, in the years preceding 1967.

The case of the demolition of the White Rock copper, silver and lead works was slightly unusual—this was one of the oldest industrial sites in the valley, and had been occupied almost continuously from 1737 to 1928. Its many buildings, chimneys and a curious arrangement of arches were found to be of considerable interest to industrial archaeologists. By 1963 many of the buildings were in such a bad state of repair that an approach was made to the Royal Commission on Ancient Monuments to see if they could be preserved. After an inspection, however, it was decided that the buildings were unsafe, and they were consequently demolished. Reclamation work was undertaken by the 48th Squadron RE (TA) in 1963.

One further demolition, that of the Llansamlet copper and arsenic works, was carried out by the 42nd Lancashire Divisional District RE (TA) in 1964, but this was linked with other Project work and is discussed later in this chapter.

Grant Aid and Major Reclamation Schemes

By 1981 the city council had become the chief landowner in the valley, and had implemented many reclamation schemes. Grant aid in respect of land acquisition had been provided by the Welsh Office and, after local government reorganisation in 1974, by the Welsh Development Agency. Up to March 1979, Swansea council had received £580 000 from the Agency

All possible sources of funding for the reclamation work were investigated by the Swansea authorities, including the EEC. The European Regional Development Fund was established in 1975 to help to rectify the principal regional imbalances within the community. As a relatively deprived region, the Lower

Swansea Valley is eligible to receive aid, and Swansea council submits regular applications through the Welsh Office, which endorses them, and then passes them on to the EEC. If the grant is approved, the money is paid to the UK government, who pass on the money to the city council. Up to 1979 grants received from the Regional Development Fund for infrastructure improvement schemes had amounted to over £1.4 million.

Some reclamation schemes carried out since 1974 have been completed without grant aid, and in its determination to ensure that progress made in the period 1974–78 is maintained, in 1978 the council approved a three-year capital budget for reclamation schemes in the valley estimated to cost a further £3 million.

Among the many improvement schemes carried out, the five described below illustrate some of the difficulties encountered in the reclamation of such a badly polluted and derelict landscape.

Removal of tip waste and use of the site by new industries

The huge Morfa tip on the west bank of the River Tawe had been produced by constant tipping from the Hafod and Morfa works since the early nineteenth century. When these works closed down in the 1920s the tip remained as an ugly reminder of the valley's industrial past. Between 1970 and 1976, four hectares of land were cleared by the removal of 63 000 cubic metres of tip material, and this was then used as foundations for roads elsewhere in Swansea, and in particular, for part of the M4 motorway which was then being extended through South Wales.

Once the tip had been cleared, the riverside site—a part of which was already occupied by light industry—was then suitable for the further development of new industry. In line with the new planning proposals, industries coming to the valley had to provide adequate landscaping around their factories to avoid the creation of the once-popular concrete industrial estate, so developing an 'industrial park'. In line with this idea, provision was made for a riverside walk which will eventually run the full three miles (4.8 km) of the lower valley.

Removal of a tip and use of the site for a new school

The Pentre-Hafod tip, composed of slag from the Vivian complex at Hafod, covered five hectares and was reputed to be one of the largest copper waste tips in Wales. In 1972–73, about 112 000 cubic metres of waste were removed and taken to another part of the city to fill in marshland which was then reclaimed and levelled for use as sports pitches.

The removal of tip waste in the valley was not a simple matter. The Project had proved that it was impossible to reduce the massive blocks of congealed waste into pieces of workable size by explosion; the only suitable method was found to be repetitive hammering by a concrete block dropping from a crane. This method was very time-consuming, but eventually all the tip material was removed from the Pentre-Hafod site at a cost of just over £400 000.

Within three years a new school, the Pentre-Hafod comprehensive school was built, and was opened in September 1976. The surrounding area was landscaped and the whole site can now be seen for miles as a new area of green in the city where once had been an ugly tip.

Removal of a tip and revegetation

The White Rock works at Foxhole had been in operation from 1737 to the 1920s, during which time hot copper slag had been continuously dumped into what had once been an attractive valley on the side of Kilvey Hill. The tip could be seen as an ugly scar on the landscape, and

Hafod and its copper tip before it was removed in 1973.
(LSVP 1963)

The same street in Hafod after the removal of the tip, and the new Pentre-Hafod Comprehensive School.
(SJL 1979)

the haphazard removal of material from it for hardcore had only served to make the scene even more desolate.

The removal of the tip, which was the first scheme started in the reclamation of the valley after the Project in 1967, served two useful purposes. In the first place, after acquiring the 33 hectare site and removing the 183 000 cubic metres of waste, a scheme was put into operation to revegetate the land which had been under the tip for so many years. Secondly, the tip material was removed to Morriston where it was used to raise and level the land on the east bank of the river. This site was later occupied by Morganite Electrical Carbon Ltd.

In preparing the White Rock site for revegetation, it was regraded, water courses were laid and the area was landscaped to give a more gentle slope to the hillside. During the time the copper waste had lain upon this site, its immense weight had compacted the underlying soil, creating conditions that would seriously affect the establishment of plants. In addition, a more serious problem caused by the tip was that the heavy rainfall had leached the toxic salts of copper, zinc and lead into the subsoil, resulting in unusually high concentrations of these metals. The prevailing acidity of the soil made the copper and zinc more readily available for uptake by plants (these metals are more soluble at low pH) and resulted in the inhibited growth of grasses. The pH value of tips and contaminated soils is a very important factor when considering the establishment of plants upon them.

In order to prevent the copper and zinc being so readily available to plants, the whole area was covered with a layer of lime which, being alkaline, raised the pH of the soil, so rendering the metals insoluble. Following the liming of the hillside, the site was then sprayed with a

The White Rock tip site after clearance and landscaping. (SJL 1978)

View over the White Rock works after demolition and clearance. In the background are the derelict ruins of the Vivians' engine sheds. (SJL 1978)

Success of early tree-planting experiments.
(LSVP 1966)

mixture of chicken manure and grass seed. The first effect of the organic constituents of the manure was to reduce the toxic effects of the copper and zinc by forming complexes with these metals, again rendering them unavailable to plants. In addition, the manure was a very good fertiliser which ensured the strong, early growth of the grass seedlings.

During the first two seasons, the grass developed well and the area was covered with a fine green sward, but unfortunately the lime and chicken manure was gradually washed out of the soil. The copper and zinc therefore returned to solution and once again became available for uptake by the grass. The consequence was that after two years of very encouraging results the grass died back leaving a sparse covering of vegetation. The spraying operation was repeated, and by 1980 grasses had established themselves again, to a certain extent, and a number of trees such as pine, birch and alder had been planted in the area. It is hoped that as the humus layer builds up gradually, so the grass cover will improve.

The tree growth on the White Rock site has been very disappointing, however; many ten-year-old trees are less than seven feet tall. The polluted and compacted soils will obviously have a serious effect on the reclamation plans for the area for a long time to come.

In addition to the reclamation work, the council leased 70 hectares of land on Kilvey Hill to the Forestry Commission. At first the Commission were reluctant to commit themselves to establishing forest areas in such a densely populated area. The plantations would be completely within the city of Swansea on the east side, and surrounded on three sides by housing estates. The area had originally been covered with birch and oak woodland, but had been stripped of all its trees by the pollution in the industrial phase of the valley. Considerable enthusiasm from the city council, backed by the Project Report, finally convinced the Commission to plant 250 000 trees on the hill. It is expected that the

View across some young plantations in the centre of the valley.
(LSVP 1965)

Commission will treat Kilvey Hill as a commercial forest, eventually producing crops of timber.

The trees planted on the hill have (by 1980) already produced a considerable improvement in its appearance, and it now forms part of the urban forest, and links in with other wooded areas planted during the 1960s. As well as being a working forest for the Commission, Kilvey Hill is starting to be of great value as a backdrop to the development of the eastern part of the city. The use of the hill by school children for educational studies, and by adults for recreation, is gradually increasing, and there is obviously great potential for the area.

Reclamation of a derelict site using new topsoil

The reclamation of the site of the Llansamlet copper and arsenic works was carried out by the Army, Project staff and volunteers. It was the only site where a complete reclamation scheme was carried out during the Project's lifetime.

As already described, a unit of the Territorial Army, the 42nd Lancashire Division District RE, demolished the buildings, flues and existing chimney stacks on the 1.6 hectare site during May and June 1964. In order to immobilise the toxic copper and arsenic still present in the tips on the area, ground limestone was spread over

The Llansamlet copper and arsenic works in 1962, 1966 and 1979.
(LSVP 1962 and 1966, SJL 1979)

91

the rubble at 13 tonnes per hectare. Contractors then brought in large amounts of soil (a poor, acid, sandy loam with a very low level of nutrients) from a nearby housing development site, and this was laid over the site to a depth of 15–30 cm. In the spring of 1965 the soil was fertilised with lime and potassium, and was then sown with a grass/clover seed mixture. Further applications of fertiliser during the following growing seasons, together with grass cutting for three years, have helped to develop the species on the site. In 1966 birch trees were planted on a small mound in the middle of the site formed by the demolished buildings. Nothing further was done to the site, but by 1980 over 25 different plant species were observed to be growing on the reclaimed site. The development and diversification of the flora indicates the successful treatment of this area.

Grass trials and afforestation

The land in the valley not utilised by smelters and factories or covered over by tips, had still been seriously affected by airborne pollutants. As discussed in Chapter 2, sulphur dioxide from the smelters resulted in the destruction of all the birch and oak woodlands in the Lower Swansea Valley, and the sparse covering of grass contributed to the scene of complete desolation. The Project staff decided to carry out a series of grass trials and experimental tree plantations which would eventually complement the effects of site clearance. It was hoped that by introducing trees and grass back into the area, people would see the valley in a very different light. As stated in the Project Report, 'The visual improvement of derelict land is an essential first step towards its eventual redevelopment.'

Grass trials. Many experimental plantings were made on each type of industrial waste and soil in the valley ' . . . so that whatever land-use plan is adopted, the findings of the botanical enquiry should be relevant to the resulting

Sparse grasses growing directly on copper waste. (SJL 1978)

Early tree planting on derelict sites by students.
(LSVP 1963)

revegetation programme.' Whilst it was possible to establish a cover of vegetation on all types of material, further work was still needed to be done to see whether the initial establishment of plant species would be followed by satisfactory long-term performance.

The two main methods used to establish an initial cover of vegetation were as follows. (i) By incorporating lime, general fertilisers, and a cheap, readily available form of organic matter such as domestic refuse or sewage sludge, a variety of plants grew, but after two and a half years the growth of shrubs and grasses declined. Further study is required to determine whether this poor growth was a result of lack of nutrients or a response to the effect of toxic metals. (ii) Some plants had managed to grow naturally on the copper, zinc and lead tip material, and these species (common bent grass and creeping bent grass) were found to have evolved a special degree of tolerance to the toxic metals. The growth of these tolerant species was encouraged by the addition of a general fertiliser and an organic amendment. The seeds from these strains were harvested, and when sown on other types of tip waste were also found to be tolerant. Work on the mechanics of metal tolerance in plants is at present being investigated in various university botany departments throughout Britain.

The botanists' study report recommended that the best treatment for the tips, if they cannot be removed, is to cover them completely with domestic refuse, for instance, then foundry sand, and then to plant into this layer. Obviously this sort of revegetation would not, by the very nature of the tip material, stand up to any heavy recreational use, but there are three very important reasons for attempting to establish plant species directly into waste material. Firstly, if the site is green and attractive, it is much more likely to be taken up by a developer than if it looks derelict and ugly. Secondly, if the ultimate use of the tip site is unknown, it is sensible to use relatively cheap methods to make it look presentable (such as direct seeding with tolerant species) rather than spending large amounts of money on regrading and importing topsoil to provide what might be only a temporary vegetation cover. Thirdly, by covering tip material with plants, the amount of wind erosion will be reduced as the roots bind the soil more securely. This will also result in the stabilisation of steep tips and help to prevent further erosion.

Early tree planting by school children.
(LSVP 1963)

Afforestation. The afforestation of the bare eroded hillsides has had a considerable effect on the valley and its inhabitants. The Project set up a number of experimental sites to discover which species of trees would be most suited to the conditions of the valley, and also to demonstrate the visual effects of revegetation on an area that had been almost devoid of vegetation for over a hundred years.

The tree planting up to 1966 took place either in large blocks by agencies such as the Forestry Commission, or in small areas by the Project's Conservator, who enlisted the assistance of local volunteers. Based on the early results of the trials, the principal species were Lodgepole

Successful plantation of trees in the centre of the valley in 1965.
(LSVP 1965)

View through woodlands towards Pluck Lake.
(SJL 1979)

pine, Corsican pine, Japanese larch, Norway spruce, birch and alder. When the Project was completed in 1966, more than 100 000 trees had been planted on sixteen sites covering a total of 20 hectares, but in the years following the first tree-planting experiments (1962 and 1963) vandalism resulted in damage to many young trees.

The Conservator and School Involvement

> To the schoolboy, a fence represents a challenge, something to break through or climb over; but trees planted with his own hands are something to care for.
>
> *Garth Christian*

The Project staff decided that the way to combat vandalism to the plantations was not fencing or patrolling the forests, but by recruiting local people, particularly youngsters, to plant trees themselves. It was hoped that this involvement would instil into the children a feeling of pride in their efforts and they would regard the forest areas as their own, to be protected rather than used as a target for vandalism. Once the forests could be established, their use by local people for recreational activities was likely to ensure their future success.

It was in 1962 that the first approach was made by the Project staff to the then Director of Education who gave his permission to allow the staff to give lectures to children during school hours to inform them of the valley's industrial history, and the reasons behind the afforestation programme.

In 1966 a full-time Forest Officer (later to become the Conservator) was appointed, and was responsible for forest maintenance and visits to local schools. Originally, four junior and four secondary schools were involved, representing almost 2000 children. Unfortunately, from the Project funds the employment of a Forest Officer could only be guaranteed until the end of 1967, even though it was agreed that such an officer was essential. The Project Report published in 1967 recommended that: 'Unless children are made partners in such schemes at an early age, the work done will almost certainly be wasted. The education programme and the maintenance of the afforested areas in the Lower Swansea Valley therefore needs to be continued on a more permanent basis.'

Successful plantations. (SJL 1979)

A plantation now established on once-derelict railway sidings.
(SJL 1979)

During the years of the Project, many close links had been forged between the city council and the University, and in late 1967 funds for the employment of a Conservator for the valley were allocated by the council to the University. The grant, apart from providing the salary, also allocated funds for the purchase of a suitable vehicle and certain equipment. On recommendation that the post should continue, further annual grants were provided until 1978. Annual contracts, whilst renewable, encouraged each Conservator to move on using this position in the Lower Swansea Valley as a stepping stone to a permanent post. Up to 1978, seven graduates of forestry or botany had been employed as Conservator, but in that year a three-year contract was offered, which allowed him to plan further ahead and also to obtain assistance from the Manpower Services Commission. In 1979 and 1980 the Commission approved two projects for the Conservator to employ a teacher, a research worker and a forest worker/warden, full-time in the valley under the Special Temporary Employment Programme (STEP). A carpenter was employed in 1979 to build an exhibition at Swansea Museum, which is described later in this chapter.

A plantation on eroded soils in the centre of the valley after 16 years' growth, and Pluck Lake.
(SJL 1979)

As the council acquired and planted more land in the valley, the Conservator's work developed into that of an educationalist rather than just a forester. By 1980, about 450 000 trees had been planted in the Lower Swansea Valley, including 250 000 trees on areas of Kilvey Hill leased to the Forestry Commission. At this time the Conservator was working closely with eleven junior schools and three secondary schools, representing over 5000 children directly involved in the reclamation of the valley.

Development of nature trails and resource material

As interest in the reclamation of the area developed through illustrated talks and films shown in schools and colleges, the demand for more information for teachers increased. A nature trail was laid to take children through the derelict parts of the valley as well as to the reclaimed areas; this led to the production of a simple duplicated information leaflet in 1964 entitled *Excursion to the Lower Swansea Valley* which was intended for use by all groups, both children and adults, visiting the valley. Following the publication of the Lower Swansea Valley Project Report in 1967, and other articles in journals, magazines and books, groups from outside Swansea have become interested in the valley. People from all over Britain have requested visits to the Project area involving studies of biology, forestry, geography, history, environmental science and sociology, and the Conservator became the person who dealt with these visitors. As schools and colleges realised the value of derelict and reclaimed land as important study areas, further resource material was essential to assist the groups studying the area.

During European Architectural Heritage Year (1975), Swansea organised its own programme to interest local adults in their urban surrounds. Using funds from the Swansea Heritage Committee, the earlier leaflet,

Birch trees growing on eroded soils.
(SJL 1979)

Damage to the leading shoot of a lodgepole pine by the pine-shoot moth. This is now a major problem affecting young trees in the valley.
(SJL 1979)

Excursion to the Lower Swansea Valley, was revised, rewritten and published as a nature trail guide *Discovering the Lower Swansea Valley* (1975, Welsh translation 1977). This was followed shortly afterwards by another booklet *Industrial Archaeology Trail—Lower Swansea Valley* (1975), which was researched by the South West Wales Industrial Archaeology Society.

Both booklets were informative and valuable to adults wishing to know more about the area, but teachers commented that they were too detailed for junior school children. With assistance from the local schools, the Library Resource Service, teachers and the Conservator, a pamphlet was produced specifically for younger children, entitled *Nature Reclaimed* (1978). This follows the original nature trail, but rather than explaining in detail each site visited, it asks questions and has space for the children to write or draw in the answers. This 'response trail' allows the children freedom to make their own decisions on a variety of topics presented in the booklet, including reclamation, transport, pollution and tree recognition. The success of this style of booklet had led

Renovated pithead buildings of Scott's Pit, now an industrial archaeological monument. (SJL 1980)

Children on nature trails. (SJL 1979)

to a new series of *Study Guides* dealing with the individual topics introduced in the *Nature Reclaimed* leaflet. Each *Study Guide* is devised by an expert with design assistance from the Resource Centre and teachers, and the production has constantly relied on close cooperation between councils, schools and the University College. By 1980 a leaflet on the study of bird life in the valley and a second on how to recognise the different trees were both in use, and studies of other topics (such as fungi and geology) were in preparation.

Educational visits by local secondary school children have involved more detailed studies of wildlife in this reclaimed landscape. One group produced a study entitled *Wild Flowers in the Lower Swansea Valley* which was published in 1976. The children investigated a number of sites in the valley and made species lists of the plants found—it was interesting to see the variety of plants which had colonised the industrial waste land. Whilst the species list of eighty plants was by no means complete, it did indicate that plants were gradually recolonising the valley.

A second report concerned a school's study of the fungi of the valley. Over thirty species have been observed, and although this is not a large number, the fact that groups of school children can be taken into the Project area to see the different types of mushroom and toadstool, for instance, is of great importance. Children who walk the nature trail during September and October are usually lucky enough to see species of the genus *Boletus* (some 50 cm tall) and the strange fly agaric with its poisonous red and white spotted cap.

A pamphlet entitled *Birds of the Lower Swansea Valley* was published in 1976 after a group of young ornithologists watched and noted the birds in the valley over the course of a year. The leaflet shows how some birds have survived through the pollution and dereliction, and have even thrived on the lack of competition, whilst newcomers are able to exploit new food supplies

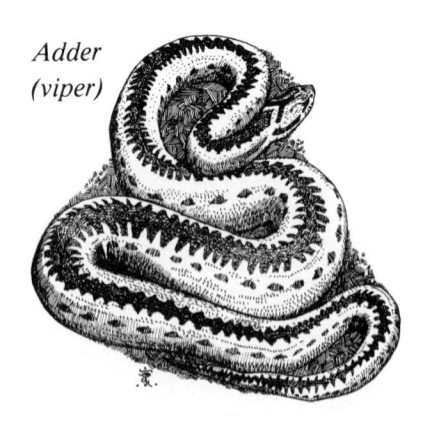

Kestrel

Boletus

Adder (viper)

Wildlife now found in the Lower Swansea Valley.
(CJM)

and fill vacant ecological niches as the vegetation of the valley develops again.

Examples of the types of bird which may have survived the period of dereliction are the kestrel, a bird resident and breeding at a number of sites; and the robin, which is probably the most abundant species. Another species to have come into the valley after the clearance of the tips and planting of trees is the coal tit, which is now resident and even more common than the blue tit. The introduction of coniferous woodland has provided food and shelter for several species of birds. The kingfisher is now taking advantage of the reclaimed landscape, and is being sighted more frequently near the river and at the small lake in the valley. The river is gradually being cleared and many species of fish are now finding it possible to live there. The lake, called the Pluck, was

Pine-shoot moth and its effects

Fly agaric

Pamphlets compiled by local school children published in 1976.

101

stocked with roach in 1962, and has thus provided another source of food for the kingfisher and the heron, both of which have been observed feeding there.

Throughout 1979 a monthly magazine was produced, containing articles about the wildlife to be seen throughout the year, as well as topical articles and features about the valley's history and personalities. The magazine—*Valley Monthly*—was produced to satisfy the demand from school children who were becoming even more concerned about the valley. By providing quiz pages and puzzles the magazine combined educational and recreational needs, and has proved to be another valuable resource for teachers.

The continuing visits to the valley by schools and colleges from all over Britain motivated the Conservator to encourage the production of further resource material. In association with the city council, a facts sheet was published in 1978, which dealt mainly with the planning statements and the progress being made with reclamation projects.

The city council also sponsored the making of a film about the history of the Lower Swansea Valley which was produced for a conference in 1979. The conference, organised through the University College of Swansea, in association with the city and county councils, looked into the changes that have taken place since 1967, and the prospects for future development. It was financed to a large extent by the Nuffield Foundation, who had helped to sponsor the original Project in 1961. The conference brought together a number of the original members of the Project staff, as well as the people responsible for studies and practical work today, and valuable ideas and information were exchanged. A book of the edited papers from the conference was published by the University College entitled *Dealing with Dereliction* (1979).

Throughout 1978 the University College and the city and county councils sponsored the building of an exhibition which was set up in Swansea Museum using photographic material to illustrate the theme of man's effect on the landscape of the Lower Swansea Valley. Research for the material took eighteen months, and the exhibition took twelve months to design and build. The framework of the exhibition was made by a carpenter employed on the Special Temporary Employment Programme, and the detailed layout of each board was produced by a technician based at the Museum, financed by West Glamorgan County Council under the supervision of the Conservator. The exhibition was formally opened by Lord Sandford and the Mayor Councillor Susan Jones in May 1979. A response trail has been produced to assist young children to obtain the full benefit of the educational value of the exhibition.

In 1978 a slide book entitled *The Reclamation of a Polluted and Derelict Landscape* was published to illustrate the variety of techniques used to reclaim derelict sites for specific purposes. Throughout the late 1970s the local Industrial Archaeology Society continued to research sites, and two further leaflets were published by the city council. The first, *Morris Town Trail,* describes the part of the city originally developed by the Morris family and later became known as the 'tinman's suburb' in the early 1900s. The second, more recent leaflet, *Swansea's Industrial Villages—Coal Mining Trail,* describes the area around the mining villages of Birchgrove and Llansamlet. All of the trail guides are packed with valuable information and are available to any interested groups wishing to visit these sites. At the same time, all the trails and resource material produced attempt to impress upon the visitor that the word 'derelict' need not be synonymous with 'devoid of life', but show that there is a great variety of wildlife to be found in this area.

Many schools throughout Britain visit field study centres to investigate the ecology of seashores, moorland, dunes or marshes. Gradually, the effect of man on his environment is also becoming a major feature of the school syllabus, and so studies of inner cities, shopping centres and derelict land are increasing.

With the development of groups such as the Ecological Parks Trust, which encourages the use of cities and derelict sites (at least on a temporary basis) for the study of wildlife, the Lower Swansea Valley would be an excellent place to establish an urban study centre.

Community Involvement

Learning about the valley, however, is not limited to work and study in school. In the early days of the Project, the Reverend E Hunt, then Vicar of Glantawe Parish, encouraged local youngsters to become involved in small work projects during their spare time. This involvement demonstrated that the young people were indeed willing to spend their own time and energy on the valley. From 1968 to 1976, when the Conservators were employed on an annual basis, the Reverend Hunt was invaluable in providing continuity by introducing each new Conservator to the area and, most important of all, to the young people.

In 1976, shortly before leaving the Glantawe Parish for a different part of the city, the Reverend Hunt opened a youth centre to provide much-needed accommodation for the youth groups organised through St Margaret's Church. In that year the Conservator did not move on, however, and was able to carry on working with the children and draw the attention of the new vicar to their valuable work.

Tree-planting sessions, simple management tasks and fire-fighting duties were carried out by members of the youth club, and in as early as 1970 the value of their work was recognised by the grant of a Prince of Wales' Award. The Prince of Wales' Committee encourages and organises environmental projects for local volunteers in Wales in order to make people more aware of their surroundings and to attract young people to voluntary work. Each year since 1970, when the Committee was established, a number of awards have been made by Prince Charles to projects which are deemed to have made an outstanding contribution to, or have increased the understanding of the environment in Wales. Such an award was a great boost to the efforts of the young people in the valley. In 1979 the tree-planting carried out by children from twelve junior schools was entered in the Shell 'Better Britain' competition sponsored by the Nature Conservancy Council and the Civic Trust. The project achieved reasonable success by being placed joint third in the regional finals (Wales, the South West and the Midlands), winning a prize of £25 and a magnificent plaque.

Following on from youth club projects, work camps were organised for overseas students who worked during their summer holidays on projects such as the construction of a small dam and windbreaks, planting trees, and even picking stones from derelict sites ready to be sown

Members of an international work camp constructing a dam on Kilvey Hill.
(*South Wales Evening Post*, 1974)

with grass. It was eventually decided that the local youngsters would also benefit if such summer schemes could be organised for them.

In 1975 an Urban Aid grant made it possible for the Conservator to organise the first adventure holiday scheme for local youngsters. The adventure holiday involved working on small projects in the valley for three days of each week and going on visits on the other two days. The work included building footpaths, digging steps, litter-picking, fire patrol, fertilising trees, cutting firebreaks, as well as many other projects. These two-day visits have taken the youngsters to beaches on the Gower, walking in the Brecon Beacons, strawberry picking and even on an army assault course. In the evening, hired feature films are shown and discos arranged.

The effect of the holiday scheme has been remarkable, and in 1976 it also won a Prince of Wales' Award. The subsequent publicity enabled many similar holiday schemes to be organised all over the Principality through the field officers of the Prince of Wales' Committee. Since 1975 yearly applications for grant aid to organise holiday schemes have been successful, and further schemes are planned for the future. Additional financial aid for the organisation of volunteer projects have been granted over the years by the city and county councils, the Prince of Wales' Committee, the Queen's Silver Jubilee Appeal, and by local organisations and industry. In 1977 and 1978 the local Carnival Committee raised a great deal of money which was then put back into local projects, and over £200 has been received from this group for use on the holiday scheme.

The enthusiasm of the children has also resulted in increased adult involvement in environmental improvement schemes. For example, regulars at a local public house, the Halfway Inn at Winch Wen, have become closely involved with the work. This public house is often used by urban and field study groups, where they are

Windbreak built by international work camp members in 1966.
(SJL 1979)

Fire-beaters in preparation for the unavoidable fires which occasionally affect the valley plantations.
(LSVP 1966)

able to meet and talk with the Conservator and local people about the valley and the changes which have occurred there. In 1978 a group within the pub, the 'Hatters', organised money-raising events; as their interest in the work of the young people increased, so they wished to be involved further. Some members of the group have assisted in tree-planting and fire-fighting, while others have worked voluntarily as supervisors on the adventure holidays.

In March 1978 the Hatters donated money to the Conservator to organise a formal club for the young people of the valley. With support from the local Carnival Committee and local industry the Lower Swansea Valley Rangers group was born.

News reports of the formation of the Lower Swansea Valley Rangers for young people. (South Wales Evening Post, March 1978)

Valley Rangers club formed

By CHRIS PEREGRINE

ABOUT 50 youngsters turned out for the initial meeting of the Swansea Valley Rangers at St. Margaret's Church Hall, Bonymaen.

The Rangers is a club for boys, girls and adults who want to improve the state of the Swansea Valley. The first meeting was organised around a disco by Steve Lavender, project conservator, who has high hopes for the club.

"All the children will receive badges, membership cards and a plastic wallet. This will ensure that the children feel part of an organisation," he explained.

A flying start to the Rangers was ensured with a cheque of £500 from the Halfway Hatters, a club formed from customers at the Halfway Inn, Winchwen.

Making the presentation on behalf of the Hatters was Swansea and Welsh rugby international, Geoff Wheel.

The Rangers are unusual in that they are supported by members of a pub club, rather than the more traditional charity organisations. The Hatters welcome the challenge of raising money for the good of the community and would like to see other pubs perform a similar function.

Now the Rangers are founded, they can embark on protecting all the trees and wildlife in the valley and help those in need, such as old age pensioners, throughout the valley.

By 1980, the Hatters had donated £650 in their continuing support of the group, and there were over 300 Rangers from the valley schools. The organisation of such a large, enthusiastic group of young people inevitably raised problems, however, and when asked to support litter-picks, tree-planting sessions, etc, there would often be a massive turnout which needs a great deal of organisation and cooperation. Consequently the Rangers group now needs further consideration, but the size of the membership demonstrates the need for more activities and facilities for the young people, who have such a tremendous amount of energy, and upon whom the future of the reclaimed areas of land depends. It is important to allow the youngsters to become involved in this work so that their energies may be used in this helpful way rather than in vandalism or destruction. In October 1979, the concept of the Lower Swansea Valley Rangers won a Prince of Wales' Award (the third for various projects involving the young people of the valley) and the Award was given to the Hatters, in the

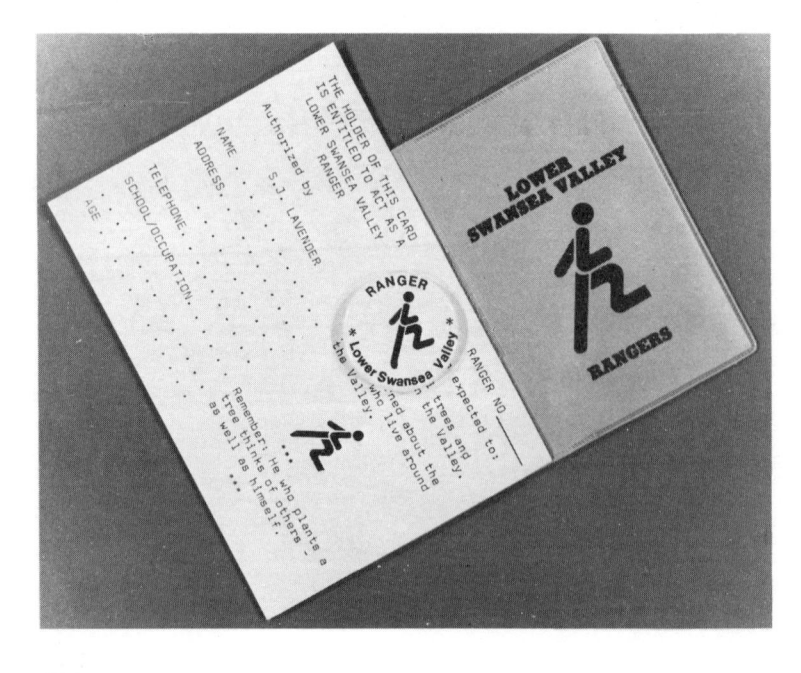

Ranger's badge, wallet and membership card.
(LSVP 1979)

hope that other pubs and local groups might be encouraged to sponsor youth organisations.

One of the most important lessons that has been learned from the Lower Swansea Valley Project is the value of close links between the young people and the Conservators, who have been able to guide the energy of the youngsters into valuable project work.

School children planting trees during winter on Kilvey Hill. (*South Wales Evening Post,* 1977)

In 1979, the Social and Community Planning Research Group carried out a qualitative evaluation of the 'Personal benefits and satisfactions derived from participation in urban wildlife projects' on behalf of the Nature Conservancy Council. The research involved visits to four sites in Britain to evaluate how and in what way the urban dweller benefits from contact with wildlife areas. The chosen sites, at Moseley Bog, Aston; Mexborough; Portland Walk, London; and the Lower Swansea Valley, differed in many ways. The conclusions from this research make it clear that both children and adults have benefited from their involvement with project work and, moreover, that this benefit was found to be greatest among those people who were directly involved with conservation work.

The success of the tree planting in the Lower Swansea Valley has depended to a very large extent on continuing education and community involvement. The vandalism of trees and wildlife is now minimal, but this has been achieved only after a long struggle against apathy, and required great enthusiasm, energy and patience on behalf of those involved. The continuing involvement of young people in the changes taking place in their own environment will ensure that the future of the valley's forests is secure.

Chapter 5
The Future

Land is a diminishing resource. Since the war we have been using about 40 000 acres of farm land every year in England and Wales. With new motorways, a rising population and the intensified working of minerals, the pressures on clean land will increase tremendously. As a nation we need to learn to husband our resources and re-use them. We are beginning to do this with water; we must also do it with land. The motto of the Lower Swansea Valley Project could well be 'New Land for Old'.
    K J Hilton, Geographical Magazine, *September 1964*

The vision of the future is of a valley in which people will live and play as well as work, without these activities intruding much on each other; a valley which offers the seclusion of wooded slopes and quiet paths as well as the activities of its workshops.
    K J Hilton, Lower Swansea Valley Project Report, 1967

The Achievements

By 1980 the population of the City of Swansea was approaching 170 000 and was still showing slight annual increases. The reclamation work being carried out in the valley was well under way, and it will only be financial restrictions which will prevent the whole area being brought back into full use by 1985.

In 1980, the twentieth year since the start of the Project, an assessment of the changes which have taken place during this time indicates the enormous amount of

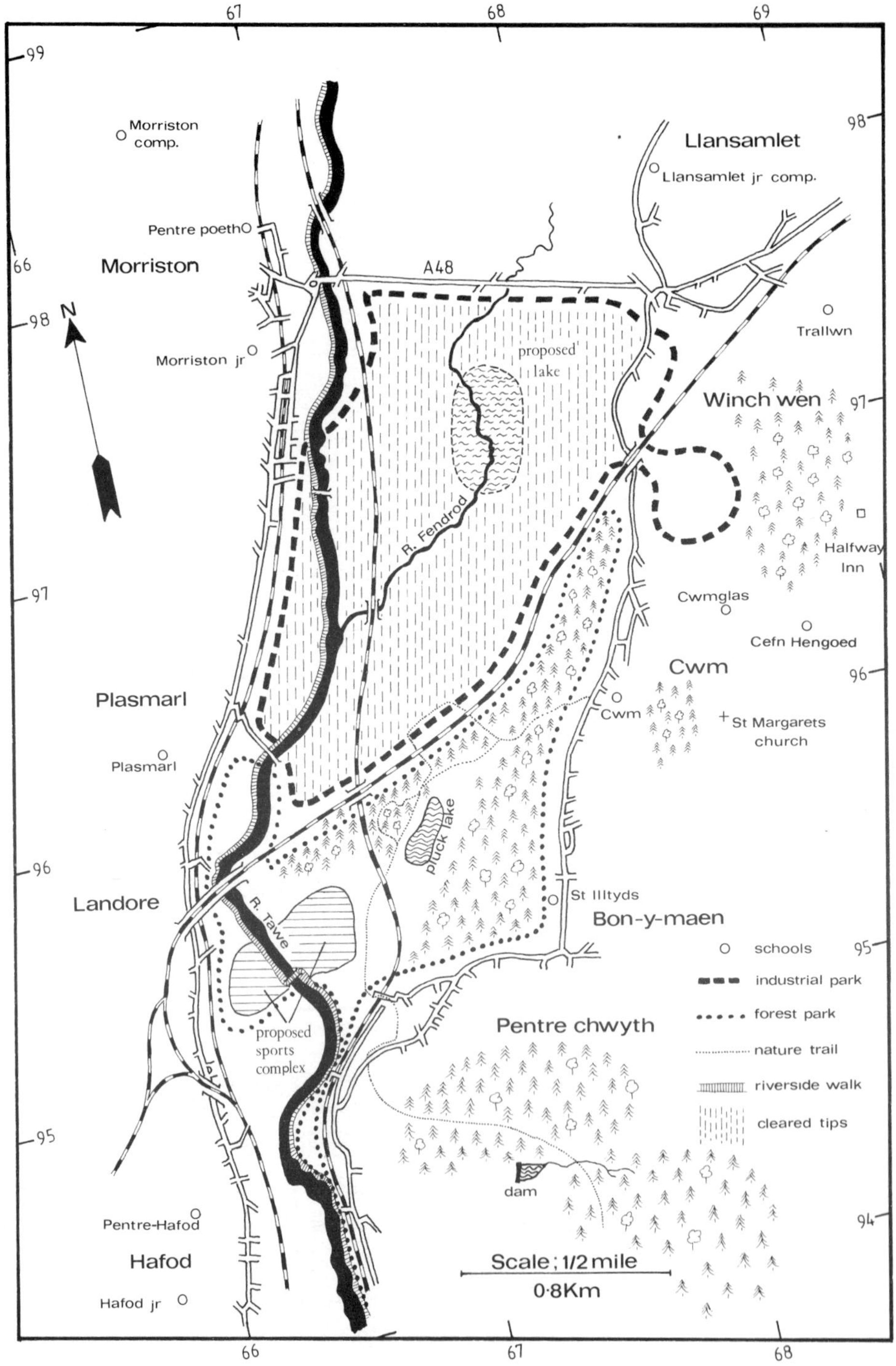
Morriston comp.
Llansamlet
Llansamlet jr comp.
Pentre poeth
Morriston
A48
Morriston jr
proposed lake
Winch wen
Trallwn
R. Fendrod
Halfway Inn
Cwmglas
Cefn Hengoed
Cwm
Plasmarl
Cwm
St Margarets church
Plasmarl
duck lake
Landore
R. Tawe
St Illtyds
Bon-y-maen
proposed sports complex
Pentre chwyth
dam
Pentre-Hafod
Hafod
Hafod jr
Scale; 1/2 mile
0·8Km
schools
industrial park
forest park
nature trail
riverside walk
cleared tips
N

Map of the Lower Swansea Valley in 1980 showing land use proposals. The forest park, the industrial park, proposed lake and sports complex, and the schools involved with the Project are illustrated.
(CJM)

work that has been completed and the degree to which the Project's recommendations have been successful.

In 1961 there were a number of reasons why the land itself was a serious inhibitor to any easy reclamation solutions, but many of these problems have now been overcome. For instance, the multiple ownership of land in the valley was a major drawback, but this has been solved by the city council's acquisition of over 300 hectares during the last twenty years, so that all the land requiring reclamation work is now either owned by or is leased by the local authority. The division of the valley by the river and the railway line is still a problem, although the closure of one line from Upper Bank to the old Swansea Vale works, and the filling in of both Smith's Canal and the Swansea Canal, have helped to ease this situation. There is still a need for improved road access to many parts of the valley, and a new road development is planned to link the M4 motorway with the city centre; part of this (the Plasmarl bypass) was opened in June 1980. As industrial sites are prepared in the area to the north of the London to Swansea railway line, so roads and services are being provided to meet their needs.

The problem of flooding in the valley still remains, but part of the plan for the development of an industrial park (to be discussed later in this chapter) includes the provision of a lake for flood relief.

Most of the seven million tonnes of waste had been removed from the valley by 1980, and the land released has been used in a variety of ways, as already discussed. For example, the site of the Morfa tip is now a light industrial estate; the site of the Hafod tip is now occupied by the Pentre-Hafod comprehensive school, and the site of the old White Rock tip has become part of Kilvey Hill forest park. Many thousands of trees have been planted on the poor, eroded soils of the valley, and these trees, since they are protected from vandalism by the local people, are now growing well.

The western side of the valley in 1968 looking south, showing a disused railway line and the Swansea Canal.
(Photograph courtesy of G Humphrys, 1968)

The same view in 1980. The Swansea Canal has been filled in and the Plasmarl by-pass has been built.
(SJL 1980)

The city council's interim planning statement for the Lower Swansea Valley (1975) proposed several major redevelopment plans, and the three-year programme approved by the Welsh Development Agency in 1977 stimulated a great drive to complete the reclamation of the valley as soon as possible. The aim of the planning statement was to draw together all the previous ideas, plans and recommendations and to produce a final report which would provide the framework for th redevelopment of the entire area.

Work in Progress

The 1975 planning statement divided the Lower Swansea Valley, from the M4 motorway in the north to the mouth of the River Tawe at Swansea Bay in the south, into six detailed planning areas. By 1980 a number of amendments had been made to this statement, however, to create five large parks within the valley. From south to north, these are:

(1) South Dock and maritime quarter;
(2) North Dock city park;
(3) sports complex/forest park;
(4) enterprise zone industrial park; and
(5) the riverside park, which will link in with the other four parks.

The development of these parks by the city council has stimulated further national and international interest in the valley. Particular interest has been shown by the

The cleared site of the Swansea Vale works being prepared for new light industries.
(CJM 1979)

115

Existing and future industrial sites will be screened as the forest plantations become established. (SJL 1979)

European Campaign for Urban Renaissance, sponsored by the Council of Europe, which was launched in 1980 with the major objective of stimulating the improvement of urban areas in Europe, and the lives of their citizens, to create 'a better life in towns.' The overall programme will concentrate on five principal themes, and in Britain five demonstration sites were selected, one of which was the Lower Swansea Valley. Each of the five themes of the campaign can be illustrated by different aspects of the redevelopment work, and these are outlined in the descriptions of the proposed parks in the project area.

(1) *South Dock and maritime quarter*

The developments in the South Dock area and maritime quarter of Swansea illustrate the first of the themes promoted by the Urban Renaissance Campaign: that of the revitalisation of existing but run-down old areas and buildings.

The South Dock and the maritime quarter first developed as a residential area, and by 1820 had become the social and cultural centre of Swansea. As the need for improved dock facilities to serve the industries further up the valley increased, however, so the maritime quarter gradually became the commercial and business centre of the city by 1880. As the years passed, however, trade declined, and so this area was badly affected. Many of the buildings became disused, and were to suffer from years of neglect.

Luckily, many of these elegant old buildings are still intact, and there are now proposals to renovate them and to bring this area of Swansea back into full use. For example, a theatre (the Dylan Thomas Theatre) has been created out of a once derelict garage area; an arts workshop has been opened in a former seamen's chapel; and a row of Georgian houses has been renovated to produce flats for almost 100 people.

The South Dock itself is to be cleared, and a new channel cut to reopen the entrance from the river. Here a marina will be developed to provide recreational facilities for people with small sailing craft, who could use the relative calm of the dock to learn sailing techniques before venturing into the rougher waters of Swansea Bay.

Many of the old derelict industrial buildings around the South Dock will be cleared, but one large warehouse has been renovated and now houses the city's Industrial and Maritime Museum. Once the South Dock has been entirely cleared it will house several floating exhibits of the museum, including a lightship, the *Helwick*, and a tug, the *Canning*.

(2) *The North Dock city park*

A second theme of the campaign is illustrated by the North Dock city park, where the city council is improving the environmental quality of an area which developed around a section of the North Dock. As in the case with the South Dock area, with the decline of the valley's industries, so the North Dock was also filled in. The filled in area was not suitable for intensive redevelopment, particularly buildings, and so it is proposed to create an urban park between the present commercial developments and to provide pedestrian routes from the Eastside through to the city centre.

(3) *Sports complex/forest park*

Most of the area of Hafod and Landore, which extends across the centre of the lower valley (see map, p 112), was covered with either derelict buildings or just bare sterile soils, and presented an awful sight to people entering the city by train. It was planned that this area should eventually be transformed into a place where residents of the Eastside, who had been deprived of accessible parkland and open space, could visit and find some peace and quiet. The 1967 Project Report also suggested that the forest areas could be extended to link in with Kilvey Hill to form some 100 hectares which could be replanted and used as recreational land.

The achievement of community participation in reclamation schemes is a third aspect promoted by the Urban Renaissance Campaign.

By 1980 the forested areas were established and the trees were growing well; many other parts of the valley had also been planted, including Kilvey Hill in the early 1970s. The success of the afforested areas has to a large extent been due to the involvement of local children through the conservators who had been based at the University College since 1967. The Project Report had suggested 'that the education programme and the maintenance of the afforested areas in the Lower Swansea Valley should be continued on a more permanent basis'. In July 1981, following a series of contracted posts at the University College, the Conservator of the Lower Swansea Valley became an officer of the city council's environment department.

Aerial photograph looking east over the Lower Swansea Valley in 1980, showing the sports complex, the Pluck Lake and some established plantations.
(© Roger David, 1980)

It is essential for the Lower Swansea Valley that a team with responsibility for environmental management be established. With the extensive planting of trees and the use of the valley by both local and national groups as a study area, the setting up of an urban study centre would be of great benefit. The city council's 1975 interim planning statement suggested various types of land use, but central to all plans has been the development of the Pluck Lake area as a recreational centre, including a sports complex. The Project Report (1967) had suggested 'a covered games stadium with running track', and in 1975 the city's planning statement proposal noted that 'an extensive area between Hafod tip and Pluck Lake has potential for a major central city sports complex.' By 1980 these ideas had become a reality, and the first stages of the building of the complex had begun.

In the *South Wales Evening Post* of 4 June 1980 the hope was expressed that 'the Industrial Revolution ravaged Lower Swansea Valley could become the foremost recreation centre in Great Britain, and possibly Europe!'

An all-weather running track costing £½ million has been laid which is up to Commonwealth Games standard, and further development schemes costing £10 million (over five years) are planned. It is hoped that other grants from the EEC, the Welsh Development Agency, central government and the Welsh Sports Council will help to provide a 3000-seat grandstand, changing rooms and a central area for field events and soccer games. In June 1980 the mayor of Swansea, Councillor Alan Lloyd, commented that 'It is felt that after the many years that people in the area had to suffer belching chimneys and pollution that it was at least their entitlement to an area dedicated to recreation, sport and amenity.'

The dominant theme of the Urban Renaissance Campaign is the improvement of the quality of urban living. In the Lower Swansea Valley, this theme is being followed by the formulation of plans for the complete removal of the remaining tips and derelict buildings, to be carried out by 1981. Certain other areas have been designated 'housing action areas', and others as 'general improvement areas' in an attempt to upgrade the older housing of Hafod, Landore, Plasmarl and Morriston. Elsewhere, new houses (at Pentrechwyth) and schools (the Pentre-Hafod comprehensive and St Illtyds junior schools) have been built, in line with the general theme of the campaign.

(4) *Enterprise zone industrial park*

The creation of an 'enterprise zone' illustrates a further aim of the Urban Renaissance Campaign, to provide social, cultural and economic opportunities in an area which has been badly affected by industrial decline. The enterprise zone is physically divided into two distinct regions by the A48 trunk road, and up to 1980 most of the development effort had been concentrated on the area to the south of this road. In 1967 it was suggested that about 150 hectares of this land should be reclaimed and developed to attract light industry, to take full advantage of the site. These advantages then included:

(*a*) its proximity to a large service centre (Swansea);

(*b*) its proximity to a large port, railway line and proposed liner train terminus;

(*c*) its position close to the A48 South Wales road from which there is easy access to the M5 and M4 motorways (when this is extended into West Wales).

The site does have its disadvantages, however, principally its physical state, even though the majority of the tips and derelict buildings have been removed since the Project Report was published in 1967. The provision of a flood relief lake is now under way, and as well as

Morganite Electrical Carbon Ltd, one of the first new industries to move into the valley following the Project Report and reclamation in 1966.
(CJM 1979)

relieving the longer-term drainage problem, the lake will also be used as a recreational facility. The M4 has already been extended into Swansea, and on into West Wales, and the city's 1975 planning statement proposed the creation of an 'imaginative and sensitive industrial environment based upon industrial park rather than industrial estate principles'.

The designation of this part of the valley as an 'enterprise zone' in 1981 will help to attract a wide variety of interests to the newly prepared sites.

The few heavy industries remaining in the valley would be expected to improve their visual appearance, and 'a footpath system could provide safe and easy access within and between groups of industries and link factories to open spaces'. The derelict sites have now all been cleared, and only a few tips from the Swansea Vale works remain. The development of the Winch Wen site in the north-eastern corner of the project area (a proposed industrial site in the 1975 planning statement), was begun, but in 1977 it was suggested that because of its situation away from the main body of industrial units, the site should be used as a residential area instead. A number of years passed, with the final decision as to the ultimate use of the site held in abeyance. The 1975 recommendations were eventually implemented, however, and the site is now a thriving light industrial estate.

In the main enterprise zone industrial park the basic services have already been laid out. Existing woodlands planted in the 1960s have survived remarkably well, despite the heavy engineering traffic, and are to be maintained and extended, with such a design to penetrate and flow through the industrial units. When these units are eventually taken up, it is expected that a high standard of design will be attained, and to achieve this a development brief produced by the city council will be available for the new developers.

A number of the peripheral industrial units are already occupied by a wide variety of concerns, such as a garden superstore, a furniture warehouse, a timber merchant, and so on. These new activities are very different from those of the last century.

(5) *The riverside park*

The proposals for the riverside park link all four previous plans and illustrate the final theme of the Urban Renaissance Campaign—to increase the involvement of the local authority. As its contribution, Swansea City Council has stepped up the development of the riverside park to allow people access for fishing, sight-seeing and walking.

Prior to these developments by the council, however, the Project had noted in 1967 that the waters of the River Tawe had shown a considerable improvement in quality; sea trout and salmon were being caught there, in the river which had once been described as an 'open sewer'. It was suggested that by building a barrage on the river, a 3.5 mile (5.6 km) stretch of sheltered water could be created which would be invaluable for providing water-linked sports facilities such as swimming, boating and fishing. 'A relatively inexpensive solution would be a barrage of a type . . . which is in essence a large fabric tube filled with water . . . As part of the improvement of the river a serious attempt should be made to clean up the banks and to provide a riverside walk from Morriston along the western bank to the New Cut bridge.'

The idea of the barrage is still only a proposal at the time of writing, but it is agreed that by maintaining a high water level the environment of the river would be vastly improved.

The Future

The Lower Swansea Valley of the 1980s bears little resemblance to that of a century ago, and although the changes that have taken place are striking, nevertheless many of the older local people now feel that something of the excitement of the valley has gone with the removal of the tips and old industrial buildings.

Public attitudes to the environment underwent a dramatic change in the 1960s, and in response to the increasing demand to clean up derelict urban areas, the reclamation work was begun. Throughout Britain large areas of land have been made derelict over a long period of time—in some cases by the dumping of household waste or abandoned cars, while elsewhere the indiscriminate disposal of industrial refuse has created the

Landscaping on the banks of the River Tawe at the New Cut. (CJM 1979)

problems. All land should be regarded as a valuable national resource, and even those areas which have been polluted and despoiled by unchecked industrial development must not be ignored. Any inner city or town can be revitalised by the reclamation of these wastelands.

The people of Swansea took the initiative twenty years ago, using the resources available to them at the time, and with the aid of many volunteers have helped to create an entirely new landscape out of the dereliction. The Lower Swansea Valley Project has been a unique venture. It was hoped that not only would the work help to put new life into a very badly neglected area of South Wales, but also that it would open the eyes of people elsewhere, both in Britain and abroad, to the fact that with a great deal of effort something can be done to make ugly, derelict land both attractive and useful again.

Riverside walk, 1980.
(SJL 1980)

This reclaimed land should then be utilised to the full by the whole community. In particular, industries, when looking for new development sites, should use this land as a first alternative, rather than 'green-field' sites which have hitherto occupied large areas of valuable agricultural and recreational land on the outskirts of cities.

But the impetus for this action must come from the people of the cities themselves, as it did in the case of Swansea, and this in turn can only come as a result of education. A derelict landscape can be used as a valuable educational resource. In 1980 the Ecological Parks Trust was established in order to promote, amongst other things, the use of neglected and derelict areas of land for educational studies. The Lower Swansea Valley has now become an excellent example of how such an area can be reclaimed, and also how it can be used by schools and universities for the study of industrial history and urban development, as well as the re-establishment of flora and fauna in a reclaimed environment.

There are no quick or easy solutions to the problems involved in the reclamation of derelict land; it is a long and slow process, fraught with difficulties which at times may appear overwhelming and almost insurmountable. The task of revitalising the Lower Swansea Valley, once described as Britain's most concentrated area of industrial dereliction, is by no means complete, but it is now well under way as a result of the initiative of one man who succeeded in convincing an entire city of the value of his plans. By making use of the lessons learned and experience gained from the Lower Swansea Valley Project there is no reason why other enterprising individuals or groups elsewhere should not be able to turn their derelict and polluted land into useful areas for housing, for industry to create new jobs, for recreation and for education. The creation of such a heritage for future generations in Swansea is one in which the city can justifiably take pride.

Bibliography

Alexander W O 1955 A brief review of the development of the copper, zinc and brass industries in Great Britain *Murex Review* **1** (15) 389–425

Balchin W G V (ed) 1971 *Swansea and its Region* (Swansea: University College)

Banwell D F 1967 The Lower Swansea Valley Project *Town and Country Planning* (November)

Barr J 1969 *Derelict Britain* (Harmondsworth, Middlesex, Penguin)

Beynon J H and Betteridge D 1979 The rise and fall of copper—a Swansea Chronicle *Chemistry in Britain* **15** (7) 340–5

Bridges E M 1966 Restoration of eroded soils and waste materials in the Lower Swansea Valley *Chemistry and Industry* p785

Bromley R D F and Humphrys G (eds) 1979 *Dealing with Dereliction* (Swansea: University College)

Brooke E H 1944 *A Chronology of the Tinplate Works of Great Britain* (Cardiff: Lewis for the author)

Cliffe C F 1848 *The Book of South Wales, The Bristol Channel, Monmouthshire and the Wye* (G F Carrington)

Davies J M 1952 The growth of settlements in the Swansea Valley *MA Thesis* University College of Swansea

Ganwell S C 1880 *District Guide*

Gemmell R P 1977 *Colonisation of Industrial Wasteland, Studies in Biology No. 80* (London: Edward Arnold)

Grant-Francis G 1881 *The Smelting of Copper in the Swansea District* (London)

Hilton K J 1964 Restoring an industrial desert *Geographical Magazine* vol. XXXVII (September) pp372–83

—— (ed) 1967 *The Lower Swansea Valley Project* (London: Longman)

Lambert W R 1968 Some impressions of Swansea and its copper works in 1850 *Glamorgan Historian* vol. V pp206–12

Lavender S J 1978 Swansea Valley scars are healing *Geographical Magazine* vol. L (6) 366–70

—— 1980 The educational value of a derelict landscape *Nature Science in Schools* **18** (2) 34–6

Lavender S J and Wainwright S J 1978 *The Reclamation of a Polluted and Derelict Landscape* (Portsmouth: Focal Point)
Mabey R 1973 *The Unofficial Countryside* (London: Collins)
Mostyn B 1979 *Personal Benefits and Satisfaction Derived from Participation in Urban Wildlife Projects* (London: HMSO, Nature Conservancy Council)
O'Neill W 1969 Siemens steelmaking at Swansea *Metals and Materials* (August) pp312–16
Orwell G 1937 *The Road to Wigan Pier* (Harmondsworth, Middlesex: Penguin)
Pollins H 1960 The Swansea Canal *Journal of Transport History* **1** 135–54
Roberts R O 1951 Dr John Lane and the foundation of the non-ferrous metal industries in the Swansea Valley *Gower* **4** 18–24
—— 1956 Development and decline of the non-ferrous metal industries of South Wales *Transcymmrodorion Society* (reprinted 1969 in *Industrial South Wales 1750–1914* ed W E Minchington (London: Frank Cass))
—— 1957 The copper industry of Neath and Swansea *South Wales and Monmouthshire Record Society No.* 4 pp126–36
Swansea City Council (pamphlets)
　Industrial Dereliction and Progress Towards Reclamation 1973
　Discovering the Lower Swansea Valley Nature Trail 1975 (Welsh translation 1977)
　Industrial Archaeology Trail—Lower Swansea Valley 1975
　Interim Planning Statement—Lower Swansea Valley 1975
　Birds of the Lower Swansea Valley 1976
　Morris Town Trail 1976
　Wild Flowers in the Lower Swansea Valley 1976
　Lower Swansea Valley Facts Sheet 1978
　Nature Reclaimed Response Trail (compiled by West Glamorgan County Council and the Conservator) 1979
　Llansamlet and Birchgrove Coal Mining Trail 1980
Swansea County Borough 1830–1909 *Medical Officer of Health's Reports*
—— 1912 *Floreat Swansea* (G Bell)
Thomas N L 1964 *The Study of Swansea's Districts and Villages* (published privately)
Thomas T M 1967 Derelict land in South Wales *Town Planning Review* (January) pp125–41
Vivian H H 1881 *Copper Smelting—Its History and Processes* (New York: Scientific Publishing Co.)

Williams T 1854 *Report on the Copper Smoke* (Swansea: Herbert Jones)
Wood J G 1813 *The Principal Rivers of Wales*

Index

Page numbers in *italic* refer to illustrations

Aberdulais, 4
Aber tinplate works, 30
Aerial pollution, *22,* 43–9, *44,* 58, 75, 89, 100, 120
Afforestation, 73, 75, 77, 79, 89, *89,* 90, 94–6, *94–7,* 98, 101, 118, *119,* 122
Alder, 89, 95
Arsenic, 15, 34, 45, 90–1
Arsenic works (Llansamlet), 12, 34, 82, 90–1, *90–1*
Arsenical smoke, 48
Aston (Moseley Bog), 110

Barques, *23,* 23, 24
Barrage (river), 73, 79, 123
Beaufort tinplate works, 30
Bell, George, 62, 78
Bessemer and copper, 27
   and steel, 32
   converter, 32
'Better Britain' Competition, 105
Birch, 1, 15, 89, 92, 95, *98*
Birchgrove, 103
Birchgrove tinplate works, 30
Bird life, 51, 101
   of the Lower Swansea Valley (pamphlet), 100, *101*
   study guide, 100
Birmingham copper works, 8, 17
Blue tit, 101
Boletus, 100, *100*
Bonymaen, 71
Bonymaen Carnival Committee, 106, 107
Botanical studies, 56, 72, 73, *73,* 77, 94, 100, *101*
Bridgewater Canal, 20
Bristol Copper Manufactories Ltd, 29
British Mannesmann Tube Company, 33, *33*
British Museum, 77
British Steel Corporation, 60
Brunel, Isambard Kingdom, *24,* 24–6

Cambrian copper works, 7
Cambrian Pottery, 7, *8*
Cambrian zinc (spelter) works, 16
Canal, Bridgewater, 20
   development of, 20
   Smith's, *18,* 20, 26, 61, *61,* 67, 73, 74, 113
   Swansea, 11, *18,* 19, *20,* 21, *27,* 61, *62,* 67, 73, 74, 113, *114*
Champion, William, 16
Charcoal, 3, 4
Clippers, *23,* 23, 24
Coal, 2–5, 7, 13, 20, 21, 32, 35, 45, 51, 61, 63
   mines/pits, 9, *18,* 19, 26, 61, 63
   mining trail, 103
   transport of, 19, 21
Coal tit, 101
Cobalt, 11, 15, 34
Community, 40, 74, 126
Community involvement, 75, 104–10, 118
Conference (1979), 102
Conservator, 76, 94–110, 118
Copper
   alloy, 17
   decline, 16, 27, 29
   early smelting, 4, 5, *10*
   eighteenth century, 6, 7
   mines, 5
   ore barques, 22, 23, *23*
   ores, 4, 5, 11–14, 16, 23–4, 34, 49, 51, 56
   output, 9, 11, 16, 27
   smelters/works, *28*

131

Birmingham, 8, 19
Cambrian, 7
Forest, 7, 36
Hafod, 9, *9,* 11–13, *12, 26,* 34–6, 43–7, 53, 60, *62,* 85, 86, *87*
Landore, 8
Little Landore, 12
Llangyfelach, 6, *6, 10*
Llansamlet, 12, 34, 82, 84, 90–2, *90, 91*
Middle Bank, 7, *12,* 19, 35, 45
Morfa (12), *14, 53, 55, 83*
Nant-rhyd-y-Vilias, 34
Rose, 8
Upper Bank, 8, 17, 19, 26, 35, 45
White Rock, 7, *8,* 11, 34, 35, 45, *57, 58, 59,* 84–8, *88*
Williams Foster and Company, 12, 29
working conditions in, 38, 39
Ynys, 8
smelting process (Welsh process), 13–16, 27
smoke composition, 45
sulphate, 35
tip material, 14, 15, 16, 36, 49, *50,* 51–6, *52, 53, 56,* 58
toxicity to plants, 55, 88, 89, 91, 92
Copper Smoke Trial (1833), 43–8, 66
verdict, 47
Corsican pine, 95
Council of Europe, 115
County Borough Development Plan (1960), 78
Custodian of Enemy Property, 34
Cwm, 71
Cwmbwrla tinplate works, 30
Cwmfelin tinplate works, 30

Darby, Abraham, 4
Davy, Sir Humphry, 49
Dealing with Dereliction, 102
Department of Scientific and Industrial Research, 70
Derelict buildings, *52, 58,* 59, *59,* 62, 66, 81–3, *82,* 91, 92, 118
land, *52, 58,* 66–9, 74–8, 85, 98–101, 105, 121–6
Dillwyn (zinc works), demolition of, 82

Director of Education, 96
Discovering the Lower Swansea Valley (nature trail), 99
Distribution of Industry Act (1945), 64, 66
Dudley, Dud, 4
Duffryn tinplate works, 30, 31

Ebbw Vale, 31
Ecological Parks Trust, 104, 126
Education, 90, 96, 98–102, 118, 126
Director of, 96
Edwards, Williams, 36
Enterprise zone industrial park, 115, 120–3
Erosion, 51, *51, 52,* 72, 81, 94, *98,* 113, 118
European Architectural Heritage Year (1975), 98
European Campaign for Urban Renaissance, 115–20, 123
European Regional Development Fund, 84, 85, 120
Excursion to the Lower Swansea Valley (nature trail), 98
Exhibition, 102, 103

Faraday, Michael, 49
Farmers, damage to crops, 43
indictment, 43, 44
testimonies at trial, 46
Forestfach industrial estate, 64, 65
Film (Swansea Valley), 102
Firefighting, 106, *106,* 107
Fish, 101, 102, 123
Flooding, *58,* 62, 63, 113
Flood relief lake, 77, 113, 121, 122
Fly agaric, 100, *101*
Forest copper works, 7, 36
Forest Officer, 96
Forest park, 115, *116,* 118
Forestry Commission, 89, 90, 94, 98
Foxhole, 51, 86
Frosse, Ulrich, 4
Fungi, 100, *101*

Garrett articulated locomotive, *26*
General improvement areas (GIAs), 120

132

Gerstenhöffer process, 49
Glamorgan zinc works, 19
Gold, 11, 34
Grasses tolerant to tips, 89, *92*, 93, 94
Grass trials, 72, 73, *73*, 88, 92, *92*, 93
Great Western Railway, 25
Greenhouses, University College Swansea, 72
Grenfell family, *37*, 37, *38*, 39
Guildhall (Swansea), 77

Hafod copper works, 9, *9*, 11–13, *12*, *26*, 34–6, 43–4, 46–7, 53, 60, *62*, 85–6, *87*
  tip, *43*, 51, *53*, 86, *87*, 113, 119
  village, 36, 37, *37*, 42, 45, 71, *87*, 118, 120
Halfway Hatters, 107, *107*
Halfway Inn, 106, *107*
Heron, 102
Hilton, KJ, 68, 70, 76, 78, 111
Housing Action Areas, 120
Hunt, Reverend E, 104

Imperial Chemical Industries Ltd, 29, 60
Industrial and Maritime Museum, 117
Industrial Archaeology Trail, 99
Industrial Development Act (1966), 79
Industrial park, 86
Interim planning statement, 71, 113–15, 119, 122
International Voluntary Service, 83
Iron, 14, 15, 31, 32, 57
  early smelting, 3, 4
  rails, 19
  sheets, 17
  sulphates, 57

Japanese larch, 95
John, Evan, 17
Jones, Robin Huws, 69

Kestrel, *100*, 101
Kilvey Hill, 1, 39, *38*, 51, 79, 86, 89, 90, 98, *109*, 113, 118
Kingfisher, 101, 102
King's Dock, *21*, 22

Lake (flood relief), 77, 113, 121, 122
Lake, Pluck, *95*, *97*, 101, 119, *119*
Landore, 4, 5, 6, 26, *26*, *34*, 34–6, 42, *43*, 47, *55*, 118, 120
  copper works, 8
  Siemens Steel Company, 32, 33, 60
  silver works, 12, 32
  tinplate works, 30
  viaduct, 23, 26, *27*, *43*, 49, 50, 68
Land acquisition, 79, 80, 84, 113
Land use proposals, 62, 63, 75–9, *112*, 119, 121
Lane, John, 5, 6
Larch, Japanese, 95
Lead, 7, 8, 11, 34, 53
  toxicity to plants, 88
Library Resource Service, 99
Little Landore copper works, 12
Llangyfelach copper works, 6, *6*, *10*
Llansamlet copper and arsenic works, 12, 34
  demolition of, 82, 84
  reclamation of site, 90–2, *90*, *91*
Llansamlet village, 19, 35, 46, 103
Local Employment Act (1960), 66
Lodgepole pine, 94, 95, 98, *98*
  pine-shoot moth of, *98*, *101*
Lower Swansea Valley Project (Chapter 3), 68–80, 81–2, 86–7, 89, 91, 94–7, 102, 104, 109, 111, 113, 125–6
  aims, 70
  area, 71, 79
  committees, 76, 78
  coordination, 70
  Director, 70
  funding, 70
  initiative, 68, 69
  organisation, 70
  Report, 76, 78–81, 89, 92, 96–98, 118, 119, 121
  staff, 92, 95, 96, 102
  studies, 71–4, *72*, *73*, *74*
  terms of reference, 70
Lower Swansea Valley Rangers, 107, *107*, 108, *108*

M4, 79, 85, 113, 115, 121
M5, 121

133

Mannesman Tube Company, 33, *33*
   brothers, 33
Manpower Services Commission, 97
Maritime quarter, 115, 116, 117
Mexborough, 110
Middle Bank copper works, 7, *12*, 19, 35, 45
Midland tinplate works, 30
Mines Royal, 4
Ministry of Housing, 69, 70
Morfa copper works (12), *14, 53, 55, 83*
   tip, *53, 55, 83*, 113
   tip reclamation, *83*, 85, 86
Morganite Electrical Carbon Ltd, 88, *121*
Morris castle, 35, *35, 55, 83*
   family, 6, 35, 36, 103
   John and Robert, 35
   town trail, 103
Morriston, 26, 30, 36, *36, 37*, 47, *58*, 71, 87, 120, 123
   spelter demolition, 82
   tinplate works, 30
Moseley Bog, Aston, 110
Muntz, 17
   alloy, 17
Museum, Swansea, 7, 97, 103

Nant-rhyd-y-vilias, 34
Nant-y-Fendrod, 73
National Library of Wales, Aberystwyth, 77
Nature Conservancy Council, 105, 110
Nature trails, 98–100, *99*, 103
   Coal Mining Trail, 103
   Discovering the Lower Swansea Valley, 99
   Excursion to the Lower Swansea Valley, 98
   Industrial Archaeology Trail, 99
   Morris Town Trail, 103
   Nature Reclaimed, 103
Neath, 5
New Cut, 21, 123, *124*
Nickel, 11, 34
North Dock, 21
   city park, 115, 117, 118
Norway spruce, 95
Nuffield Foundation, 70, 102

Oak, 1, 89, 92
Open-hearth process, 32, 33
Ores, 4, 5
   (*see also* copper, zinc)

Penclawdd, 11
Pentrechwyth, 35, 71, 120
   cottages, *36*
Pentre-Hafod comprehensive school, 86, *87*, 113, 120
   tip, 86, *87*, 113
Photographic record, 83, 84
Picture Post, 78
Pine, Corsican, 95
   Lodgepole, 94, 95, 98, *98*
   shoot moth, *98, 101*
Plasmarl, 71, *83*, 120
   by-pass, 113, *114*
Pluck Lake, *95, 97,* 101, 119, *119*
Pollution
   aerial, *22,* 43–9, *44, 58,* 75, 89, 100, 120
   Copper Smoke Trial, 44–8
   derelict land/buildings, *52, 58*–*60, 59*–69, 74–8, 80–5, *82,* 91–2, 100–105, 118–22, 124–6
   River Tawe, 56–9, *57*
   solid tip waste, *43, 50,* 51–6, *52*–*9,* 62, 71, 72, 77, 80–6, *82, 83, 92,* 94, 113, 120, 121, 124
Population
   Morriston, 36
   Royal Commission on the Distribution of Industrial Population, 64
   Swansea, 1, 9, 35, 39–43, 48, 63, 74, 111
Portland Walk, London, 110
Port of Swansea Sanitary Authority, 40
Port Talbot, 33
Pottery, Cambrian, 7, *8*
Prince of Wales' Award, 104–7
   Committee, 104–6
   Dock, *21,* 23

Queen's Dock, *21,* 23
Queen's Silver Jubilee, 106

Railway
  development in valley, 26-7, *26,* 34
  early national development of, 21, 24
  first in valley, 25-6, *26*
  lines, 67, 73, 113, *114*
  London-Swansea line, 49, 50, 68, 69, 79, 82, 113
Rangers, Lower Swansea Valley, 107, *107*
Reclamation, 66, 69, 76, 79, 111, 114, 125-6
  early plans, 62-4
  early programme, 65
  schemes, 80 (Chapter 4)
Resource material, 98-103
Response trail, 99, 100, 103
Revegetation, 56, 87
  techniques, 72, 73, 77, 88, 89, 92-4
Richard Thomas and Baldwin, 60
  and Company, 31
River Tawe, 1, 5, 7, *8,* 11, 13, 21-4, 26, 35, *53,* 56-9, *57,* 62, 67, 73, 78, 85, 88, 101, 113, 115, 117, 123, *124, 125*
  barrage, 73, 79, 123
  riverside park, 115, 125
  riverside walk, 75, 86, 123, *124,* 125, *125*
Roach, 102
Rocket (Stephenson's), 24
Rose copper works, 8
Royal Commission on Ancient Monuments, 84
Royal Commission on Distribution of Industrial Population, 64

Salmon, 123
School children, 83, 90, *94,* 98, *99,* 102, 109, *109,* 118
School involvement, 74, 75, 95-104, *99,* 120, 126
Studies, 98, 99, 100, 103, 104, 106
Scott's pit, *99*
Sea trout, 123
Ships, 1, 5, *5,* 23-4, 40, *57*
Shipping, 9, 11, 21, 22, 24, 40, *61*
Siemens, C W, 32, *32,* 33
  derelict works, *59*
  old laboratory, *32, 34*
  open-hearth process, 32

Steel Company, Landore, 32, 33, 60
Silver, 7, 11, 12, 13, 14, 15, 34
  Landore works, 12, 32
Singleton Abbey, 37, 38, *39,* 69
Singleton Park, 37, *38, 39,* 69
  Swiss cottage in, *39*
Smelters, copper (*see* copper smelters)
  zinc (*see* zinc smelters)
  metal ore, 3, 4, 9, 16, 19, 21, 27, *28,* 29, 34, 38, 42-8, 51, 62, 92
Smelting techniques (*see* copper-smelting process)
  early copper, 4, 5
  metal ore, 3, 4, 13
Smith's Canal, *18,* 20, 26, 61, *61,* 67, 73, 74, 113
  Charles, 19
  family, 35
  John, 19
Social and Community Planning Research Group, 110
South Dock, 21, *22*
  and maritime quarter, 115, 116, 117
South West Wales Industrial Archaeology Society, 99, 103
Special Areas (Development and Improvement) Act, 63, 64
Special Temporary Employment Programme, 97, 103
Spelter works (*see* zinc works)
Sports complex, 75, 76, 115, 118, 119, *119,* 120
  water sports facilities, 123
Spruce, Norway, 45
St Illtyds Junior School, 120
St Thomas, *38,* 39, 71
Stephenson, 24
  Rocket, 24
Steel, 29, 31-4, 43, 51, 53
  early production, 4
Stewart and Lloyds, *33,* 34
Study groups, 106, 107, 119
Study guides, 100
  wild flowers, 100, *101*
  birds, 100, *101*
Sturtevant, Simon, 4
Sulphur, 13, 34, 45, 48, 49
  dioxide, 92

135

Sulphuric acid, 11, 35, 45, 49, 56
  smoke, 49
Summer holiday schemes, 105, *105,* 106
Swansea, 1941, 65, *65*
  area, *2*
  bay, 1, *5,* 115
  borough, 6
  borough council, 62, 69, 70
  Canal, 11, *18,* 19, *20,* 21, *27,* 61, *62,* 67, 73, 74, 113, *114*
  Castle, 1, *3*
  city, *22,* 41, 47, 63, 65, *65,* 118, 126
  city council, 64, 78–85, 97–9, 102, 106, 113, 116–19, 122
  docks, 12, 21–4, *21,* 25, 61
    King's, 22, *21,* 22
    North, *21,* 22
    Prince of Wales', *21,* 22
    Queen's, *21,* 22
    redevelopment of, 65, 11
    South, *21,* 22, 11
  early, 1
  East, 37, 74, 89, 90, 118
  Guildhall, 77
  harbour, *22, 23*
  Harbour Trust, 22
  Heritage Committee, 98
  mayor of, 103, 120
  Museum, 7, 97, 103
  Royal Jubilee Metal Exchange, 31
  University College, 69, 70, 73, 77, 78, 97, 102, 118
  Vale Railway Company, 26
  Vale works, 19, 30, *44,* 59, 113, 122
    cleared site of, *115*
  Valley (film), 102
  village, 1, 2, 3, 7, *7*
  West, 12, 37, *38, 39,* 69, 74

Territorial Army, 81, 82, *83,* 84, 90, 91
Ticketing, 13
Tin, 15

Tinplate, 29, 30, 31, 51, 57, 67
  departments of work, 30
  mill teams, 30, 31, *40, 41*
  workers, *40, 41*
  works, *28*
    Aber, 30
    Beaufort, 30
    Cwmbwrla, 30
    Cwmfelin, 30
    Duffryn, 30, 31
    Landore, 30
    Midland, 30
    Morriston, 30
    Upper Forest, 30
    Worcester, 30
Tips, *50,* 51–6, *52–9,* 62, 66, 71, 72, 77, 80–5, *82, 83, 87, 92,* 94, 113, 120–4
  removal of, 85–8, 120
  resmelting of, 34
  uses for, 62, 85–7
Townsend, Chauncey, 7, 8, 16, 19
  mines, 19
  wagonway, 19
Transport (in the valley), *18,* 19–26, *20, 26, 27,* 56, 63
Tree planting, 72–7, 83, 89–96, *89, 93, 94, 96, 97,* 104–8, 113
  trials, 94
Trustees, 22–3

Unemployment, 63, 64, 66
University College, Swansea, 69, 70, 73, 77, 78, 97, 103, 118
  greenhouses, 72
Upper Bank copper works, 8, 17, 19, 26, 35, 45
  zinc, 17
  railway, 26, 113
Upper Forest tinplate works, 30
Urban study centre, 104, 119

Vandalism, 75, 95, 96, 108, 110, 113
Valley Monthly Magazine, 102
Villiers' zinc works, 17

Vivian engine sheds, *26, 88*
   family, 7, 9, *9,* 11, 12, 17, 29, 36, 37, *38, 39,* 44–8, 69, 86
   Hafod works (*see* copper smelters, Hafod or Hafod copper works)
   Henry Hussey, 11, *11,* 46, 47, 49
   John, 11
   John Henry, 11
   Richard Hussey, 11
Voluntary work, 81, 83, 91, 94, 95, 104–10, *105, 106*
   aims of, 95, 96

Wagonway, Townsend's, 19
Water sports facilities, 123
Weaver building, *61*
Welsh Development Agency, 80, 84, 114, 120
Welsh process (of copper smelting), 13–16, 27
Welsh Office, 69, 70, 80, 84, 85
Welsh Sports Council, 120
West Glamorgan County Council, 99, 100, 102, 103, 106
White Rock copper works, 7, *8,* 11, 34, 35, 45, *57, 58, 59,* 84, 86, *88*
   demolition of, 84
   reclamation of, 86, 87, 88, *88,* 89
   tip, 51, *54, 59, 82,* 113
Wigan, 55, 59

Wild flowers, 51
   of the Lower Swansea Valley (pamphlet), 100, *101*
Williams Foster and Company, 12, 29
Winch Wen, 71, 122
Worcester tinplate works, 30
Working conditions in smelters, 14, 38, 39

Ynys zinc smelter, 8
Yorkshire Imperial Metals, 59

Zinc chloride, 35
   decline of works, 29
   derelict works, *60*
   ores, 34
   output, 17, 29
   oxide, 35
   spelter works, 16–19, *28,* 30, 43, 51, 53, 59, *60*
      Cambrian/Dillwyn, 16, 82
      Glamorgan, 19
      Morriston, 82
      Swansea Vale, 19, 30, *44,* 59, 113, *115,* 122
      Upper Bank, 17
      Villiers', 17
      working conditions in, 19
   tip waste, 51–6
   toxicity to plants, 55, 88, 89, 93